Iron,
Nature's
Universal
Element

Iron,
Nature's Universal Element

**WHY PEOPLE NEED IRON
& ANIMALS MAKE MAGNETS**

*Eugenie Vorburger Mielczarek
and Sharon Bertsch McGrayne*

Rutgers University Press
NEW BRUNSWICK, NEW JERSEY,
AND LONDON

Library of Congress Cataloging-in-Publication Data
Mielczarek, Eugenie V.
    Iron, nature's universal element : why people need iron & ani-
mals make magnets / Eugenie Vorburger Mielczarek and Sharon
Bertsch McGrayne.
        p. cm.
    Includes bibliographical references and index.
    ISBN 0-8135-2831-3 (alk. paper)
    1. Iron—Physiological effect.   2. Iron in the body.   3. Bio-
magnetism.   4. Magnetoreception.   I. McGrayne, Sharon
Bertsch.   II. Title.

QP535.F4 M54 2000
612'.01524—dc21
99-056540

British Cataloging-in-Publication data for this book is available from
the British Library.

Manufactured in the United States of America

THIS BOOK IS DEDICATED TO SCIENTISTS
WHO SPEND THEIR LIVES DISPELLING MYSTERIES

AND TO
JOHN, MARY, AND TERRY

AND TO
GEORGE, FRED, RUTH ANN, AND TIM.

CONTENTS

LIST OF FIGURES   ix

PREFACE   xi

ACKNOWLEDGMENTS   xv

Introduction: A LEGACY OF IRON   1

1   What Was Iron Doing at Life's Birth?:
LIFE WITHOUT OXYGEN   5

2   Catastrophe: THE ARRIVAL OF OXYGEN   25

3   Grabbing and Storing: CONTROLLING IRON   31

4   The Smallest Living Magnets:
AVOIDING OXYGEN   49

5   Hemoglobin and Myoglobin:
HARNESSING OXYGEN   71

6   Migrating Animals: MAGNETIC TRAVEL   105

7   Iron and the Planet's Ecosystem:
SEAS AND SOILS   131

8   Feeding the World's Poor:
IRON DEFICIENCY   149

GLOSSARY   161

BIBLIOGRAPHY   177

INDEX   197

FIGURES

I.1. Earth's anaerobic beginnings 2

1.1. The iron atom 8

1.2. Tube worms 14

1.3. Godzilla 18

1.4. Carl Woese's tree of life 22

2.1. The evolution of life on Earth in relation to the origin of the solar system 26

3.1. Enterobactin, *E. coli*'s siderophore 34

3.2. A bacterial cell 35

3.3. A gated porin channel 36

3.4. A biological scale 37

3.5. A ferritin iron-storage molecule 41

3.6. A cell releases iron from transferrin by using endocytosis 43

4.1. A magnetic bacterium 54

4.2. Magnetic domains and domain walls 58

4.3. Earth's magnetic field 61

4.4. Northern and Southern Hemisphere magnetic bacteria 62

4.5. Bacteria that produce magnetite without becoming magnetic 64

4.6. The four electrons responsible for iron's magnetism 67

4.7. Magnetic domains 68

4.8. Electrons on neighboring atoms aligned by the exchange interaction   69

5.1. Myoglobin's structure   75

5.2. Hemoglobin's structure   76

5.3. The heme   77

5.4. Docking   80

5.5. Hemoglobin delivers oxygen   82

5.6. Hemoglobin compared to myoglobin   83

5.7. Myoglobin delivers oxygen   84

5.8. Maternal hemoglobin delivers oxygen to fetal hemoglobin   89

5.9. The development of hemoglobin   90

5.10. Sickle cell logjams   98

6.1. Earth's magnetic field   112

6.2. A homing pigeon with its battery pack   116

6.3. The North Atlantic gyre   123

6.4. A baby sea turtle wearing its Lycra swimsuit   124

7.1. Toles cartoon on iron seeding   136

7.2. An overview of nitrogen fixation   141

7.3. The courtship across kingdoms   143

7.4. The chemistry of nitrogen fixation   144

PREFACE

Iron, Nature's Universal Element is about iron's role in living organisms ranging from bacteria and plants to people and other animals. Iron played an important role in the origin of life on an anaerobic, iron-rich Earth. Because of this early iron-dependent biochemistry, virtually every life form on Earth today depends on this metal. Nature forged the iron on our planet into useful forms, sometimes in molecules that run essential biological processes and sometimes in minuscule magnetic compasses that help their living hosts to navigate.

Much of the material is being presented for the first time to a general readership. Although the science of iron is complex, this book describes it in nontechnical terms. Some readers will be delighted by the absence of scientific detail, but others may be disappointed. Thus the glossary at the back of the book is followed by an extensive list of primary sources for those who want more technical details.

Our present understanding of iron's function in the living world is the result of many individual scientists working in a variety of disciplines. A number of these scientists are featured in the book. Some made important discoveries while performing seemingly insignificant tasks. Others spent their entire careers working on problems that they thought were almost insurmountable. Whether they were fascinated by the migration of birds through the night sky or challenged by a professor's request to "Isolate the compound and find its crystal structure," the scientists attacked their problems with interdisciplinary relish. Their stories illustrate the lively interplay between molecular biologists, ornithologists, field biologists, physicists, oceanographers, chemists, geologists, physicians, and ecologists.

The book begins with two chapters on the formation of an iron-rich planet, the origin of life on an anaerobic Earth, and one of evolution's surprises: the appearance of molecular oxygen in the atmosphere. Recently, oceanographers and geologists, by studying the bacteria around seafloor hot springs and in rock far below Earth's surface, have taken a peek at what life on the young planet may have looked like. Iron's early abundance on the planet made the metal easily available to living systems.

Chapters 3 through 6 explore the dependence of larger life forms on iron-containing molecules and tiny magnets. Chapter 3 discusses three classes of molecules—siderophores, ferritins, and transferrins—that help living systems cope with the oxygen that flooded Earth's atmosphere. Chapter 4 deals with bacteria that assemble magnets to travel up and down along Earth's magnetic field lines and locate their particular oxygen niche. Chapter 5 describes the use of hemoglobin and myoglobin by large organisms, including humans and other mammals. The chapter also explains in simple terms the biochemistry of blood, the physiology of exercise, a number of intriguing animal hemoglobins, and medically important genetic diseases, including hemochromatosis, the thalassemias, and sickle cell disease. Chapter 6 concerns one of nature's most dramatic mysteries—the migration of birds, turtles, salmon, and other animals, migration that depends on iron magnets. The chapter details some of the exquisitely controlled animal experiments, particularly with homing pigeons and New Zealand rainbow trout, that led to the deciphering of this phenomenon.

The book's final two chapters deal with global issues: iron's influence on Earth's oceans, plant life, and human population. A sea story in chapter 7, part science and part politics, concerns controversial experiments that fertilized the ocean with iron in an attempt to alleviate the harmful build-up of carbon dioxide in the atmosphere caused by the greenhouse effect. The chapter also explains nitrogen fixation by legumes and bacteria and underscores the importance of the process for feeding the planet's growing population. On land, the interaction of legumes with bacteria in the soil manufactures nine essential amino acids that humans cannot make. Nature's ability to make these amino acids depends on a large protein with thirty iron atoms.

Chapter 8 describes the alarming consequences of low-iron diets in poor countries, including long-term cognitive damage in iron-deficient children. Women and children are most at risk for iron deficiency, especially in vegetarian societies.

Iron, one of the most familiar elements on Earth, is truly one of the "metals of life."

Eugenie Vorburger Mielczarek *Sharon Bertsch McGrayne*
Fairfax, Virginia *Seattle, Washington*

ACKNOWLEDGMENTS

This book ranges far beyond physics and biology to include geology, medicine, nutrition, chemistry, oceanography, and microbiology. In writing the book, we were of necessity heavily dependent on the goodwill and assistance of experts in other fields. Thus, while acknowledging that any mistakes remain our own, we want to thank those experts who helped us with this book.

Harold Morowitz, Krasnow Professor at George Mason University, strongly encouraged this project, and the university administration assisted in innumerable ways.

A number of scientists graciously consented to interviews with S. B. McGrayne: Kenneth P. Able, Richard P. Blakemore, Jack B. Corliss, Jack Dymond, Richard B. Frankel, Bruce W. Frost, James L. Gould, Pauline M. Harrison, Kenneth J. Lohmann, Gary J. Massoth, J. B. Neilands, Roswitha Wiltschko, Wolfgang Wiltschko, and Carl R. Woese. All quotations come from interviews with the authors, with one exception. The conversation between Ralph Wolfe and Carl Woese appeared in "Microbiology's Scarred Revolutionary," an article by Virginia Morell (*Science* 276 [1977]: 697–703).

Others reviewed and critiqued the book in its early stages: Philip Aisen, Larry L. Barton, Robert Ehrlich, Earl W. Prohofsky, Maria Taylor, and Doreen Valentine. We owe particular thanks to the scientists who reviewed and critiqued various chapters; the book is much the better for their pains. They are Kenneth P. Able, John Lea-Coxe, Richard J. Diecchio, Richard B. Frankel, Victor R. Gordeuk, Pauline M. Harrison, Jayne L. Hart, Alice Honig, David A. Hutchins, Kenneth J. Lohmann, Gary J. Massoth, James A. Metcalf, Elaine Monsen, Harold J. Morowitz, David G. Nathan, J. B. Neilands, Max F. Perutz, Albert P. Torzilli, Fernando E. Viteri, Michael M. Walker, Marianne

Wessling-Resnik, R.P.J. Williams, Roswitha Wiltschko, and Wolfgang Wiltschko.

Other specialists who kindly helped us with information, advice, or suggestions are Nancy L. Adamson, James Beard, Fred Bertsch, Ruth Ann Bertsch, Mary Bicknell, Berit Borch-Johnsen, Justin Brown, Jacob Cadity, Robert J. Celotta, Sallie W. Chisholm, John R. Delaney, Jay De Long, Jack Dymond, Philip Ekstrom, Mrs. Thomas Emery, Gary Evans, Frank A. Ferrone, Clement A. Finch, Michael Fish, Arthur Foreman, Miriam Foreman, Katherine B. Gebbie, Daniel G. Gibson III, Donald Griffin, Marvin Hass, Frank Haw, Sally Heap, Barbara A. Jones and colleagues, William Keeton, Donald Keister, Philip Klebba, Robert A. Klocke, Betsy Lozoff, Gloria B. Lubkin, Milt Martin, Cynthia McIntyre, Norman R. Pace, Tommy J. Phelps, Christine Poole, William O. Robertson, Timothy St. Pierre, G. Stamatoyannopoulos, Joanne Schuh Vorburger, Günther Wächterhäuser, Charles Walcott, Mary Welles, and Carl R. Woese.

We also owe particular thanks to our agent, Carolyn Krupp; the illustrators, Michael Hollenbach and Jeannine Harned, students at George Mason University; administrative assistants Mary A. Thomas and Heather M. Wier at George Mason University; the library staffs of George Mason University, the National Agricultural Library, and the National Institute of Health; the Cooley's Anemia Foundation, Sickle Cell Disease Center, American Society for Clinical Nutrition, and the Protein National Data Bank at Brookhaven National Laboratory; Dr. Mielczarek's students at George Mason's Learning in Retirement Institute; colleagues in the physics, chemistry, and biology departments at George Mason University who patiently answered all queries; the biochemists, physiologists, and physicians of the East Coast Iron Club; and the expert editorial eye of Robert E. L. Adamson.

Iron,
Nature's
Universal
Element

Introduction
A LEGACY OF IRON

Four billion years ago, iron was more precious than gold. It was so important for the development of early life that even today virtually every life form on Earth requires iron. Thanks to their early history, living organisms—from bacteria to people—walk a narrow line between having too much and too little of this vital element. Iron is the most abundant metal on our planet and the fourth most abundant element in Earth's crust. Its popular image of solidity and stolidity is matched by the fact that it has the most stable nucleus in the universe. Yet, paradoxically, an atom of iron can be surprisingly responsive to its physical and chemical environment. Chemistry is fundamentally the exchange of electrons between atoms, and iron is first and foremost a deft and facile electron exchanger. Slipping electrons on and off, iron plays a crucial role in brokering the electron transfers that fuel the formation of new molecules. Given iron's ubiquity and usefulness, it is not surprising that living cells evolved a chemistry that still depends heavily on the element.

Today we know that the greatest diversity of life occurs in the microbial world and that microorganisms depended on iron chemistry billions of years before humans existed and began to make iron tools and weapons. Some microbes evolved to make minuscule iron magnets. Modern biochemistry has revealed that living systems are crucially dependent on iron. Some scientists speculate that the metal may have played an important role as an energy provider in early biochemistry. Some have even hypothesized that life may have begun—ironically—on the surface of fool's gold, the iron compound known as pyrite.

One and a half billion years ago, long before the appearance of advanced multicellular life, photosynthesis began to flood Earth's at-

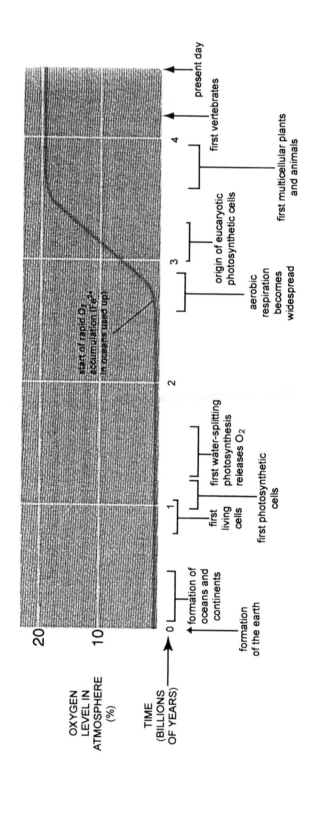

mosphere with oxygen and gradually rusted away the form of iron on which early life depended. As living organisms struggled to adapt to the new, oxygen-rich atmosphere, our planet experienced what may have been the most catastrophic extinction in its history, the loss of many species that could not tolerate oxygen, the anaerobic species (see figure I.1).

Confronted with oxygen, early life forms faced a crisis. Organisms either ceased to exist or evolved into life forms that could tolerate oxygen. Today oxygen pervades our atmosphere, and free iron has become a mere trace element in the biosphere. Nevertheless, virtually all common organisms need iron for growth and development. Iron is crucial to the energy metabolism of both aerobic and anaerobic life forms, and iron-sulfur proteins are present in all living organisms.

Within living cells, iron supports a multitude of reactions. Other metals support only a handful. Iron complexes facilitate the formation of hydrogen molecules from single hydrogen atoms; the formation of water from oxygen and hydrogen molecules and from peroxide; and biological processes such as respiration, photosynthesis, nitrogen fixation, and DNA synthesis. Some of the most famous biological molecules—hemoglobin and myoglobin—contain iron. As a result of iron's pervasive importance, most biologists have concluded that the element was obligatory for emerging life.

Living creatures today use iron in an astonishing variety of ways. Humans and other vertebrates rely on the iron atoms in red blood cells to carry oxygen from the lungs to the rest of the body. Iron strengthens the teeth of gnawing rats and gilds the livers of dugongs, the "sea cows" that inhabit shallow tropical waters from the Middle East to the Philippines. Iron is crucial to the growth of common disease-causing microorganisms such as *Salmonella* and *Escherichia coli* (*E. coli*). And some species of bacteria, birds, fish, and amphibians have even developed the uncanny ability to assemble special tiny iron magnets. They assemble magnetic iron oxide or iron sulfide particles in their

FIG. I.1. EARTH'S ANAEROBIC BEGINNINGS.
For two and a half billion years, Earth developed with only small amounts of molecular oxygen in its atmosphere. From Bruce Alberts et al., Molecular Biology of the Cell, *3d ed. (New York: Garland Publishing, 1994), figure 1-17. Reprinted by permission.*

bodies to navigate dark and murky worlds. Iron also helps fertilize our oceans and soils, and the lack of iron in protein-poor diets dooms millions of people to listlessness and impaired intellectual development.

Oddly enough, despite iron's abundance, it can be toxic. For example, millions of people accumulate toxic iron overloads because of inherited blood diseases such as the thalassemias and hemochromatosis. The latter is today considered one of the most common genetic diseases in the United States.

The story of iron on Earth begins with the role of iron atoms in the formation of our planet and moves through the world of evolving life, primitive microorganisms, and progressively more complicated iron-managing molecules to global systems for migrating animals, fertilizing the surface of Earth, and feeding its human population. From the simple to the complex, from the microscopic to the planetary scale, from the thoroughly understood to current research frontiers, this book follows iron through the biological world. We inherited our dependence on iron from Earth's iron-rich beginnings. This is the story of our legacy of iron and the scientists who are uncovering it.

1

What Was Iron Doing at Life's Birth?
LIFE WITHOUT OXYGEN

Prelude

Peering through a saucer-sized porthole at the seafloor, three scientists gaped at an iron-drenched landscape. An iron-rich hot spring was bursting from the ocean's floor, precipitating towers of iron compounds, spreading iron fluff for miles around, and feeding hosts of never-before-seen primitive creatures. How did these organisms survive, even thrive, in complete darkness and under intense pressure in an environment devoid of oxygen?

Today many biologists and geologists view these iron-filled seafloor communities as living laboratories, mimics of conditions that existed when life first evolved. We know that iron formed our planet's core and permeated its mantle and crust, but did iron also play a role in life's birth? Staring through the porthole of their research submarine in 1976, scientists could only wonder at the answer.

An Iron-Rich Planet Forms

Iron's abundance on Earth is a relic of our astrophysical history. Matter in the universe began as hydrogen about 15 billion years ago. As massive stars formed and then died in supernova explosions, nuclear energy processes in the burning stars produced heavier elements, including carbon, then silicon, and finally iron.

In the midst of all this fire and brimstone, the nuclei of iron remained stable, accumulating in massive numbers. The stability of an element depends on how much energy binds together the protons and neutrons in its nucleus. Nuclei are caught between two opposing forces:

one favoring a large nucleus and the other a small one. Iron stands at the equilibrium point between these forces, and the twenty-six protons and thirty neutrons of its most stable and common isotope, ^{56}Fe, are held together more tightly than those of any other nucleus in the universe. In the heat of the early universe, elements with nuclei lighter than iron's tended to fuse together to make heavier elements, while elements with nuclei heavier than iron tended to fission apart into lighter elements. As gravity forced a platter of whirling particles to coalesce into Earth 4.6 billion years ago, it pulled the molten iron droplets toward the center, where iron alloys formed a solid inner core and a molten outer core. In this maelstrom, Earth became an iron planet.

Earth's first billion years were a hellish baptism by fire. Bombarding the planet, meteorites and asteroids incinerated its surface. Oceans boiled, lightning flashed, torrents of acid rained down, and volcanoes erupted. The atmosphere filled with dust that blocked the sun's light and with sulfur compounds that smelled like rotten eggs. At first there was little or no free oxygen and no ozone layer to absorb the sun's ultraviolet rays, and Earth's surface was sterilized by radiation and by searing temperatures. As these global forces raged, free electrons forged and broke atomic bonds, and atoms organized and reorganized themselves in changing arrays of molecules. In this dynamic sphere, iron had no peers.

Iron is not only the most abundant metal element on our planet but also one of the most versatile electron exchangers. The nuclei of all iron atoms contain twenty-six positively charged protons and between twenty-eight and thirty-two uncharged neutrons. An electrically neutral iron atom has twenty-six negatively charged electrons to counterbalance its twenty-six positively charged protons. When iron atoms participate in chemical reactions, they usually transfer one, two, or three electrons to neighboring atoms to form compounds. In compounds, iron exists primarily in two forms, or charge states, ferrous (Fe^{++}) and ferric (Fe^{+++}). Ferrous iron (Fe^{++}) lacks two electrons because it has released them to its neighboring atoms, whereas ferric iron (Fe^{+++}) has released three electrons to its neighbors (see figure 1.1). With easy access to such a variety of charge states, iron atoms can take part in an enormous number of chemical reactions. Few elements are as flexible when it comes to helping make inorganic and organic molecules.

ELECTRON CLOUDS

Only the outer electrons in the iron atom play major roles in iron chemistry. Loosely bound to iron's nucleus, these outer electrons can easily slide over to other atoms to create iron compounds. The rest of iron's electrons are held so tightly to the nucleus that they cannot participate in chemical reactions.

Electrons occupy clouds around the atoms. Imagining these clouds is difficult if one thinks of the electrons as a point object. Instead think of the electrons as a wave. According to experimental evidence, all objects in the universe possess both wave-like and particle-like personalities. To imagine the wave-like electrons of an iron atom, mentally blur a picture of twenty-six distinct point electrons into a cloud. Quantum mechanics predicts that electrons are most likely to be found where electron clouds are densest. When two or more atoms come together to form a molecule, they sense their neighbors' clouds and rearrange their own clouds accordingly.

According to the principle of physics that governs the creation of stable molecules, atoms can continue to interact until they reach a state in which they need less energy to remain bound together as a molecule than they would need to remain apart. To break a molecule into its constituent atoms, energy must be provided by an external agent, generally by heating the molecule or by applying electrical forces. For example, two oxygen atoms in Earth's atmosphere need less energy to exist paired together as an O_2 molecule than they need to exist separately. When atoms bind together in a molecule, they share parts of their electron clouds and resist efforts to pull the clouds apart.

Iron and Developing Life

Moving beyond iron's role in forming inorganic molecules, scientists began to believe that iron also provided some of the energy needed by developing life. To the chemist, life *is* the dynamic movement of electrons from atom to atom. The cellular reactions that process molecules are the vigorous signatures of life. And while cellular chemistry has many players, including proteins, enzymes, and amino acids, it is directed by the science of energy transfer, thermodynamics.

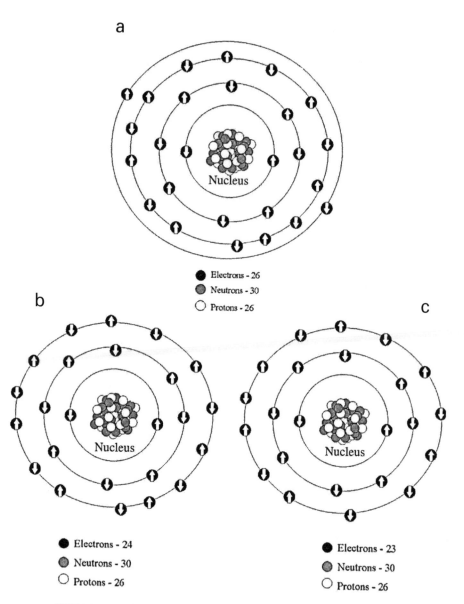

FIGURE 1.1. THE IRON ATOM IN ITS NEUTRAL *(a)*, FERROUS *(b)*, AND FERRIC *(c)* STATES.

The arrows on the electrons represent spin direction.

Atoms donate and acquire electrons in order to find the relationship that they can maintain with the least energy. As good conductors of electricity, metals do this easily. Given the abundance of iron, carbon, nitrogen, hydrogen, and sulfur on early Earth, it is not surprising that life's earliest biochemistry was driven by these chemicals. Inevitably, metals like iron that easily grab and donate electrons become the catalysts for reactions of organic compounds made with carbon, hydrogen, nitrogen, and oxygen. For example, matrices of carbon polymer chains marked much of life's early chemistry, and iron atoms spaced along the chains could transfer electrons around the matrix with alacrity. Gliding to and fro, the electrons helped build many of biology's electronic features.

Chemists refer to the transfer of electrons as reduction and oxidation reactions, or redox reactions. Reduction is the gain of electrons. Oxidation—which does not necessarily involve oxygen at all—is the loss of electrons. These so-called redox reactions are the lifeblood of cellular functioning, and iron's facile cycling of electrons made the metal an indispensable catalyst. Sliding an electron over to another atom, ferrous iron (Fe^{++}) is oxidized to ferric (Fe^{+++}) iron. Slipping an electron off another atom, ferric iron (Fe^{+++}) is reduced back to ferrous iron (Fe^{++}). The cycle can be repeated over and over. Each time iron donates an electron, it provides the energy for another chemical reaction to occur.

Before molecular oxygen invaded Earth's atmosphere, iron was plentiful in its soluble ferrous state, and ferrous iron was often called upon to donate an electron to facilitate the formation of molecules.

More recently, biochemistry and genetics have clearly demonstrated that living organisms are chemical machines, progressing from simple, unicellular forms to more complex organisms. Over millions of years, nonliving molecules gave rise to living, replicating, and self-assembling cells. And as multicellular life forms evolved, their biochemical systems also grew increasingly complex.

But what fueled the first living organisms? Later in Earth's history, photosynthesis would power living cells, turning sunlight into energy for cells to grow and function. Was there, however, an energy source to fuel living organisms before photosynthesis? Could it have involved iron and sulfur? So far, no one knows; although iron's flexibility certainly makes it a contender. In the meantime, new discoveries are challenging the leading origin-of-life theory, known as the primordial soup.

For many years, this theory was embodied in the famous experiments conducted by Stanley L. Miller in 1953. Miller was inspired by the Nobel Prize–winning chemist Harold Urey and by a book, *The Physical Basis of Life,* written by English physicist J. D. "Sage" Bernal. Urey and Bernal suggested that the organic compounds needed for life might have formed from some of the components of Earth's early atmosphere: methane, ammonia, water, and hydrogen. To test their hypothesis, Miller simulated primordial conditions with a mixture of methane (CH_4), ammonia (NH_3), and hydrogen (H_2). Then he ignited it with a lightning-like electric discharge. "The water in the flask became noticeably pink after the first day, and by the end of the week the solution was deep red and turbid," Miller observed. Analyzing the mixture, he discovered a rich soup of amino acids. He assumed he had produced amino acids under conditions that were likely to have existed on primitive Earth.

Miller's experiment was quick and simple, and soon other researchers were generating many of the small organic molecules found in present-day living cells, including amino acids and sugars and the purines and pyrimidines that are components of DNA. Scientists who carried out experiments like Miller's assumed that the compounds important in biology today were the precursors of primordial life. Indeed the organic compounds found in meteorites are quite similar to the products of these experiments.

Stanley Miller's recipe for the beginning of life remains widely accepted, and many high school biology students are still taught that it is the only biochemical theory of life's origin. But many scientists believe that other, more recent and testable theories and discoveries should also be considered. Some researchers believe that Miller's recipe will have to be entirely replaced.

Miller's experiments reflected a belief that the compounds in early life must have contained carbon, hydrogen, and nitrogen with small amounts of sulfur and phosphorus thrown in. Scientists emphasize that Earth's earliest atmosphere did not contain much molecular oxygen, and new techniques for identifying trace elements in extraordinarily large bodies of matter reveal that at least twenty elements—including iron—are essential for life. Like Charles Darwin, Miller also believed that life must have begun in shallow sunlit pools at moderate temperatures. During the late 1960s, however, microbiologists Thomas Brock of the University of Wisconsin and Jim Brierly of Montana State University found microorganisms in hot springs in Yellowstone National Park. The creatures were not only surviving, they were thriving at unheard-of temperatures: 167 degrees F (75 degrees C) and higher. Until then, biology textbooks had emphasized that life on Earth depended crucially on sunlight and photosynthesis. The discovery that hot springs could support abundant life without any sunlight at all struck a powerful blow at Miller's soup recipe. What would replace it?

Life on the Seafloor

In 1976, four years after earning his doctorate, John Corliss, a junior geochemist at Oregon State University, secured National Science Foundation funding for a series of deep-sea dives to look for hot springs on the ocean floor. Oceanographers had been speculating for several years that thermal springs might occur along the biggest geographic features on Earth, the midocean ridges that wind 40,000 miles (64,000 kilometers) around the globe like seams on a baseball. The ridges mark the boundary lines between thin plates of oceanic crust. Edging away from the ridges, these plates move as much as 8 inches (20 centimeters) yearly to the frequent accompaniment of earthquakes and volcanic eruptions.

At the time of Corliss's expedition, during the International Decade of Ocean Exploration, most scientists thought that metal deposits along the ridges had emerged with primordial water and lava from submarine volcanoes. Corliss had a different idea. He thought that seawater sank through cracks in the seafloor and was heated, whereupon it dissolved minerals from the rocks and then came back up through other fissures in the seafloor as hot springs.

During the next winter, Corliss's expedition of three research vessels, twenty-five scientists, and twenty-six technicians sailed from the coast of Ecuador in search of hot springs along the Galapagos Rift in the eastern Pacific Ocean. Taking turns in a three-person submersible vessel called the *Alvin,* they made a total of twenty-four dives to the seafloor. Most of the eight principal scientists had never dived in a submersible. Almost all of them outranked Corliss in professional experience and age. Corliss had earned his Ph.D. from Scripps Institution of Oceanography late in life, at the age of thirty-six. Before that he had studied geology, oceanography, chemistry, and the philosophy of science as a graduate student, first at the University of California at Berkeley and then at Stanford University and Scripps. Although he was forty years old at the time of the cruise, he was still an untenured assistant professor.

Nevertheless, he had written the proposal for the expedition and was its director. So accompanied by his thesis adviser, Tjeerd H. van Andel, and the *Alvin*'s pilot, Jack Donnelly, Corliss made the first dive. Crammed next to electronic equipment inside the submersible's 7-foot-wide cabin, the three men huddled together in darkness and near freezing temperatures, unable to do much more than squirm to stretch their muscles. During the 1.5-hour, 1.5-mile descent to the seafloor, little pings from acoustic transponders broke the silence.

Once on the sea bottom, the three scientists became instantly alert, peering through the *Alvin*'s 5-inch-wide ports. They cruised slowly north 3 to 6 feet (1 to 2 meters) above the sunless waste of the seafloor. They hoped to find a buoy they had lowered earlier near a possible sea-vent location. As usual, the ocean bottom was devoid of organisms; only the occasional crab or fish drifted by.

Soon Corliss noticed that the water outside *Alvin* was getting rather milky. Then a sensor sounded, announcing an increase in the water temperature outside. As Corliss fiddled with the sensor and watched the temperature gauge rise, the pilot suddenly exclaimed, "There're big clams out there!" They had discovered the first hot-spring field, a rocky area jam-packed with luxuriant life.

As the *Alvin* collected a cluster of the "clams"—later identified as giant mussels—a fountain of milky water shot up through fractures in the rocks. Apparently the mussels had been covering cracks in the rock and living by filtering microbes from the warm water. Although earlier photographs taken with a remote surface camera had suggested that bivalves might live in the area, *Alvin*'s explorations established that they lived at and off the hot springs.

The next day, when van Andel and Jack R. Dymond of Oregon State made *Alvin*'s second dive, their navigational system failed to work, and they could not find a hot spring. In the meantime, though,

Robert D. Ballard of Woods Hole Oceanographic Institution had lowered a combination camera–temperature-sensing-system into the water and saturated the seafloor with photographic runs. On the third day, February 15, at 19:09 hours, Ballard's sensor suddenly registered a spike in temperature. Later that night, when film from the camera was developed onboard, frame 19:09 revealed hundreds of mussels blanketing the rocks below.

When Corliss took another turn in *Alvin,* with John M. Edmond of the Massachusetts Institute of Technology, they located the site of the 19:09 photograph. There, a fountain of warm, milky blue water registered a comfortable 62 degrees F (17 degrees C), considerably warmer than the surrounding water. Around this vent was a lush, 50- to 80-yard-wide oasis filled with mussels as big as dinner plates. The scientists collected a large number of samples so that they could characterize the fluids in some detail and build the story of how the sea vent's circulation worked. Later that day, when the water bottles were opened in the ship's laboratory, the stench of hydrogen sulfide filled the room. For crew members, it was the sweet smell of success. They had discovered a "universe" almost totally dependent on hydrogen sulfide (H_2S) and almost devoid of oxygen. It was a world that scientists had thought was extremely hostile to life.

Later descents in *Alvin* discovered vents populated with 12-foot-tall tube worms, brown mussels, crabs scampering about like free spirits, curling worms, sea anemones, limpets, blind shrimp, and other creatures. The worms stuck brilliant blood-red tips out of white plastic-like tubes (see figure 1.2). The mussels had scarlet flesh. (Both the tips of the worms and the flesh of the mussels turned out to be rich in a special form of hemoglobin that has a high affinity for the oxygen dissolved in hot-spring water.)

Corliss was too busy running the expedition to get excited. Almost the youngest faculty member on board, he was allocating dive, equipment, and laboratory time among senior scientists. Many times he had to say no to people who had never been told no before, especially by an assistant professor. "It was a pretty high stress situation," Corliss recalled years later. "I really didn't get hit with how powerful the expedition was until the last day when suddenly all the responsibility fell off. Then I had a few hours to think of what we'd done, and that's when it was exciting. It was a beautiful day, a strong wind came up,

FIGURE 1.2. TUBE WORMS.
Reprinted with permission of Dave Brenner, Alaska Sea Grant College Program, Fairbanks Alaska, www.uaf.edu/seagrant.

and the captain of the research ship dressed the ship in flags to celebrate. . . . That's when I got excited."

As they cruised back to port, the scientists assessed their findings. They realized that the springs must be filled with mineral-eating microbes, the bottom of a biochemical food chain that does not rely on photosynthesis. Corliss and his colleagues, moreover, were already thinking of the sea vents as potential sites for the origin of life. Their speculations have long since been confirmed—all except their origin-of-life theory.

Today scientists know that hot spring creatures thrive, grow, and propagate despite utter darkness, pressures of 2 tons per square inch, and hydrogen sulfide–rich water. One-celled organisms in the water metabolized the hydrogen for energy, although most one-celled organisms use oxygen for energy. Crowding around the vents, thousands upon thousands of larger complex organisms dine on the microorganisms. So nourishing is the feast that later divers to sea vents discovered a tube worm that grows almost a yard a year, faster than any other invertebrate in the sea. In the dark elsewhere, it would grow only 6 inches yearly.

Corliss's expedition vastly expanded our understanding of ocean chemistry and our definition of life-sustaining environments, and it fed speculation among other scientists that life might not have begun in a benign primordial soup. Corliss and John Baross, a University of Washington microbiologist on the expedition, challenged the soup recipe during the 1980s. Lending weight to their arguments, A. Graham Cairns-Smith, a Scottish chemist, contended that the odds were stacked against the proper molecules' finding each other in a vast solution of primordial soup. Cairns-Smith suggested that clay might have served as a solid matrix or framework for the first self-replicating living systems to grow on. Later discoveries modified the clay hypothesis, but Cairns-Smith's interest in surface rather than solution chemistry has become widespread. Eventually, some scientists would tell themselves that if microbes could luxuriate in the harsh and extreme conditions of a seafloor hot spring without photosynthesis, they might survive in outer space too.

Several years after the discovery of the sea vents, Corliss abandoned oceanography. Others continued the exploration of sea vents on the ocean floor.

On the basis of the amount of heat that Earth's core is losing, scientists estimated that the oceans are spotted with five thousand such vents. Organisms living there produce approximately 10 million tons of biomass yearly. No one has guessed how much living matter these vents may have produced billions of years ago, when the hot sulfur-rich conditions would have been the rule rather than the exception. Presumably far more biomass would have been produced.

Today most scientists agree that vents are created when icy seawater seeps into seafloor cracks and is heated under pressure by magma to 1,100 degrees F (600 degrees C). Leaching minerals from the rocks as it expands and rises, the water shoots back up through the seafloor laden with sulfur, carbon, iron, and hydrogen and with compounds such as hydrogen sulfide, methane, and carbon dioxide.

Until the sea vents were discovered, scientists thought that all the minerals dissolved in seawater had originated in rocks on land and washed from rivers into the sea. Now geologists know that many of the minerals in seawater are dissolved by hot water inside Earth's crust and enter the oceans during volcanic eruptions. Over the course of 10 million years, all the water in the oceans circulates through the vents. At that rate, seawater has probably completed the magma-to-ocean cycle hundreds of times over the course of Earth's history.

During the last five years of the twentieth century, oceanographers came to realize a startling fact: every underwater volcanic eruption that scientists had watched close-up was accompanied by a vast outpouring of heat-loving microorganisms from below the seafloor. These microorganisms extract the minerals that they need for growth from rocks below the seafloor, and the microbes are in turn eaten by tube worms, shrimp, crabs, and other larger creatures.

The temperatures and chemical compositions of sea vents have convinced scientists that the oasis communities there are veritable nurseries for primitive organisms using chemical energy rather than light energy and feeding on inorganic compounds which contain hydrogen, sulfur, iron, and manganese. The dominant species around the vents convert methane, not oxygen, to carbon dioxide. Most creatures on land use oxygen for energy, but the sulfur-loving microorganisms that inhabit the surfaces of vent tubes and chimneys use molecular hydrogen instead. Because oxygen was scarce on early Earth, such organisms are believed to resemble ancient life forms.

Iron permeates these sea-vent communities. The plumes of water that belch up out of the vents contain 1 million times more iron than normal seawater. As the hot plumes hit cooler water, black particles—mostly the iron sulfide known as pyrite—precipitate to form smoke-like clouds. Over thousands of years, the precipitates have formed mounds the size of the Houston Astrodome. Until it collapsed, a fifteen-story tower called Godzilla, built primarily of pyrite, sat on the Juan de Fuca Ridge, 1.4 miles off the coast of Washington State (see figure 1.3). Parts of Godzilla were scorched black by heat from the vent's 626-degree-F (330-degree-C) plume of water; and cooler parts were coated furry white with microbial life.

The iron-rich water plumes that burst out of such vents can be traced 600 miles as they float horizontally over the seafloor. Their iron particles take years to gently settle, like dust motes, on the seafloor. Eventually iron carpets broad swaths of the ocean bottom near the ridges with fluff balls of iron oxyhydroxide, as soft as dust bunnies under a bed. Crumbly and friable, the barely crystallized dust forms an enormous bed of electron-rich iron surfaces where chemical reactions can occur. Iron-dependent microbes nestle and luxuriate amidst the fluff. On such metallic surfaces—perhaps on iron itself—many scientists today believe that life on Earth began. Many of the sea-vent organisms collected and cultured in laboratories are encrusted with iron minerals. One organism ringed with iron extracts the metal from the water, oxidizes it, and uses this electron transfer reaction for energy.

Archaea Discovered

Much of this enormous reservoir of microbes consists of an entirely new class of life called archaea, which Professor Carl Woese of the University of Illinois identified in 1977, a year after the discovery of Corliss's sea-vent community. Archaea are one of the most ancient forms of microbes known. Their biomass may rival that of all other life forms on Earth's surface. Tube worms at the sea vents farm archaea, much like ranchers raising cattle. Gill structures in the tube worms pull warm sea-vent water and archaea into their body cavities. There, archaea live symbiotically, extracting minerals from the water and multiplying happily. When the tube worm gets hungry, it eats the archaea. Covered with thousands upon thousands of these archaea-eating tube worms, the iron sulfide deposits on the seafloor look woolly white.

As an undergraduate physics major, Woese had been taught to divide life into five kingdoms: animals, plants, fungi, protozoa, and bacteria. Four of the five had multicellular forms. Unicellular organisms included in these branches, commonly called simply bacteria, were either procaryotes or eucaryotes. (Eucaryotes have a nucleus, and procaryotes do not.) Microbiologists tried to classify bacteria by shape

FIGURE 1.3. GODZILLA.

A fifteen-story tower built primarily of pyrite precipitated from 626 degree F (330 degree C) water flowing from a vent in the seafloor. The small figure at the lower level is the three-passenger submersible Alvin. Godzilla grew on the Juan de Fuca Ridge 1.4 miles (2.3 kilometers) off the coast of Washington State. From V. Robigou et al., Geophysical Letters 20 (1993): 1,187–1,190. Copyright by the American Geophysical Union.

and metabolism. Woese's physics background, however, convinced him that the world operates according to far deeper and simpler principles than shape and nutrition. He decided to use the new tools of molecular biology to explain the evolution of bacteria; and in 1966 he began to compare the nucleotide sequences in the nucleic acids of various bacteria.

Woese photographed the sequences of ribonucleic acids, RNAs, in microorganisms and then tried to match these snippets of genetic information with those in other microorganisms. Soon his University of Illinois office in Urbana was filled with thousands of films, carefully stored in large, yellow Kodak-film boxes. Only Woese and one or two others were even able to read the films. Gradually, over a decade, he began to understand the relationships between the various types of microorganisms. Then, in 1976, a close colleague, Ralph Wolfe, suggested that Woese look at another type of microorganism.

Wolfe, a microbiology professor at the University of Illinois, often searched sewage treatment plants, the rumens of pigs and cows, and the mud of sedimentary lakes for anaerobic organisms to study. Stirring oxygen-starved lake mud, he could bring bubbles of gas to the surface. When he caught the gas in an inverted funnel and lit a match, the gas burned blue. This "combustible air," as it used to be called, is methane. It plays an important role in the digestion of ruminants and other animals. Wolfe suggested that Woese look at these methane-making microorganisms, which depend only on inorganic material for nutrients. Eight of them had been discovered; they had different shapes but similar biochemistry, and no one knew where they fit on the evolutionary tree of life.

So Woese analyzed the genetic material from two methane-making microbes, one from an Urbana sewage-treatment plant and the other from a cow's rumen. He was stunned to see that the microbes were not bacteria at all. "They were completely missing the oligonucleotide sequences [short chains of nucleic acids] that I had come to recognize as characteristic of bacteria," Woese explained to Wolfe as they walked down the hall. "Come out of orbit," Wolfe told him. "Of course, they're bacteria; they look like bacteria." But Woese knew that nucleotide sequences were more important than outward appearance.

Within a year, Woese had collected other anaerobic microbes that grew in harsh and exotic environments. Some were salt-loving creatures that had invented photosynthesis without chlorophyll and lived in heavily saline environments like the Dead Sea, the Great Salt Lake, salt evaporation ponds, desert oases, and the like. Others were sulfur-metabolizing organisms that lived in boiling sulfurous pools. "Oh, it was a lovely time," Woese recalled. "Very heady."

Eventually, Woese built a three-branched tree of life with limbs divided among bacteria, his archaea, and the multicellular eucaryotes that include people (see figure 1.4). Woese's tree makes it clear that most of life on Earth consists of one-celled microbes. And the first living organisms on the planet may have been archaea, instead of bacteria. The oldest archaea were heat-loving and anaerobic. Bacteria and archaea branched off from one another 4.3 billion years ago.

At first, Woese's archaea were viewed as exotics from harsh, hot, and anaerobic environments. Then, in 1993, archaea were also discovered in seawater off the coasts of California, Massachusetts, Singapore, Alaska, and Antarctica: in short, around the globe. They constitute 30 percent of all extremely small plankton (less than one-millionth of a meter in diameter). Thus, archaea could be the most abundant organisms on Earth. Oceanographers studying the mid-ocean ridges came to realize that their sea-vent communities teem not with bacteria but with archaea.

Woese's discovery—long doubted and sometimes ridiculed—was confirmed in 1996, when the complete sequence of nucleic acids in the DNA of a methane-making archaea, *Methanococcus jannaschii,* was published. Oceanographers in *Alvin* had discovered this archaea in the crevices of sea vents. The sequencing of its DNA revealed the surprising fact that people are more closely related to archaea than to bacteria. Earth's first organisms may have evolved not in Miller's soup of organic molecules but in hellish boiling acidic pools like Yellowstone's hot springs or in deep-sea volcanic vents. Woese, once viewed as a renegade, was suddenly considered a visionary pioneer.

Life inside Rocks

While scientists had come to accept that life could subsist on the surface of rock, the possibility of life *inside* rock was unthinkable. Then, in 1992, an interdisciplinary team of scientists funded by the United States Department of Energy discovered microorganisms living 1.7 miles under northeastern Virginia in shale, hardened sedimentary rock. The microbes had been trapped in sediments laid down when dinosaurs lived, 230 million years ago. Since 1992, scientists have found fossils of microbes in granite under the Baltic seacoast; living microorganisms in basalt under the Columbia Plateau in eastern Washington State; and bacteria under New Mexico living off 100-million-year-old organic matter. Since granite and basalt are common and widespread, their rock-living microbes may be too. Coping with hot, anaerobic, and nutrient-short environments, one of these microbes was quickly named *Bacillus infernus.*

The oldest known fossils of bacteria have been found in Western Australia in 3.5-billion-year-old stromatolites, formed from dense mats of cyanobacteria (formerly classified as blue-green algae). Because cyanobacteria photosynthesize and produce gaseous oxygen, their fossilized remains are taken as a sign that microbes were producing molecular oxygen 3.5 billion years ago. Stromatolites became important in the growing debates about the origin of life because they formed when the planet was still being bombarded by meteorites. In this volcanic atmosphere, burgeoning life may have taken refuge beneath the seafloor or in underground rocks.

Evidence was mounting that life may have begun as simple organisms in a violent, anaerobic, iron-rich atmosphere far earlier than previously thought: within a billion years of Earth's formation. Given the fact that Europa, one of Jupiter's moons, also has a rocky volcanic center with surface water and ice, speculation started that primitive microbial life may have developed elsewhere in the solar system, too. If so, these microbes would probably also be unicellular, anaerobic, and chemically disposed to use iron and sulfur as energy sources.

A New Theory of Life's Origin

Meanwhile, Günther Wächtershäuser, a German patent lawyer interested in the chemistry of sea vents and hot springs, proposed an origin-of-life theory radically different from Stanley Miller's gentle soup. Wächtershäuser had earned a doctorate in organic chemistry from the University of Marburg in Germany. Abandoning chemistry for the law, he continued to wonder casually about the origin of life. His friends, including Carl Woese and the late English philosopher Sir Karl Popper, urged him to work out his ideas, but fearing ridicule, he did not publish them until 1988.

Wächtershäuser believed that mineral-based chemical reactions

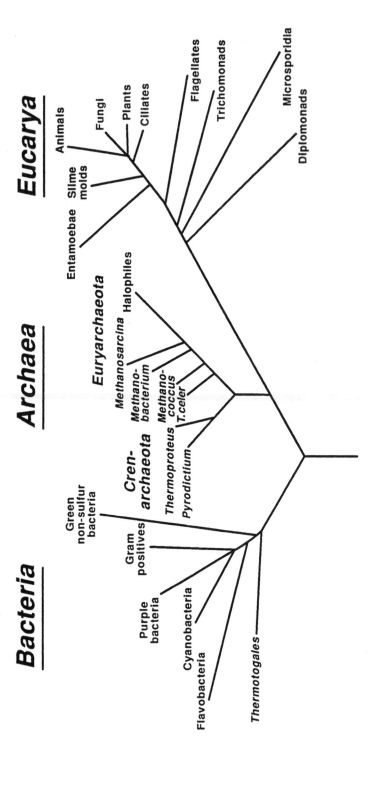

could have generated biological compounds. And he thought that chemicals floating freely in air or water would have had difficulty getting together. So, he theorized, negatively-charged biological chemicals would need a stable surface to move around on as they tried out different reactions.

Wächtershäuser hypothesized that life originated when organic molecules reacted with the hot surfaces of iron-sulfur clusters of pyrite to form iron-sulfur complexes. Iron-sulfur proteins are nature's modular, multipurpose structures. Clusters of sulfur atoms bound with one to eight iron atoms are present in all the most important biochemical reactions. They are found in mammals, plants, and microorganisms, including ancient archaea. Because iron-sulfur proteins are so widespread, many scientists believe that they are extremely ancient. But with no experimental evidence to back up his idea, few paid attention to Wächtershäuser's theory.

Then, in 1997, he and Claudia Huber, a chemist at Munich Technical University, published results of experiments that modeled a chemical reaction found in volcanic vents. They concluded that the earliest energy source for life was the formation of pyrite from iron and hydrogen sulfide in an iron-sulfur world. As they observed, iron sulfide is ubiquitous in hot sulfurous vents and volcanic settings. When a reporter questioned Stanley Miller at the University of California at San Diego, where he now works, the *New York Times* reported that Miller dismissed the experiment as "sort of blah to me, that's nothing."

Despite all the speculation about the volcanic origin of life, no one knew whether heat-loving organisms could actually have gotten the energy they needed without molecular oxygen. In 1998, groundwater microbiologist Derek R. Lovley and his colleagues proved in their University of Massachusetts laboratory that both extremely ancient heat-loving archaea and bacteria can use ferric iron (Fe^{+++}) under near boiling conditions to convert food into energy, much as higher

FIGURE 1.4. CARL WOESE'S TREE OF LIFE.
The root of the tree—the beginning of life about 4 billion years ago—is located at the conjunction of the archaea and bacteria branches. Mammals are located on one of the most remote branches on the eucarya line because they developed late in Earth's history. Courtesy of Carl R. Woese.

animals use oxygen. Iron enzymes in two primitive organisms facilitated a chemical reaction changing ferric iron to ferrous iron. For a brief instant, as electrons slipped from one iron form to another, they loaned their energy to the living organisms. Lovley chose ancient archaea and bacteria that had diverged from each other far back in time, in hopes that their common features had appeared in the earliest life forms. At least in the laboratory, ancient creatures could indeed use an iron-based biochemistry to harness the energy of an anaerobic planet.

Whether life actually began on iron compounds is not yet known. What is known is that long after the volcanism and asteroid collisions had subsided on Earth, another catastrophe struck developing life. It was a crisis of planetary proportions: the arrival of oxygen molecules. So, iron's story leaves the early planet's iron-drenched, oxygen-free biochemistry. As iron begins its journey through the biological world, we shift focus from the ultrasmall world of protons and neutrons and the early microbial life of Earth's first two and a half billion years to consider the last billion years. Early life forms will unwittingly create this catastrophe. Evolving organisms will produce a sudden and dramatic rise in the oxygen content of Earth's atmosphere and threaten their own age-old, iron-based way of life. And once again, life's ability to survive will depend on its skillful use of iron.

2

Catastrophe

THE ARRIVAL
OF OXYGEN

Life on Earth has at times been devastated by great ice ages, flip-flopping climates, volcanic eruptions, asteroidal collisions, massive extinctions, and other natural disasters, but the most catastrophic of these was the appearance of oxygen molecules in the atmosphere beginning a little less than 2 billion years ago.

Before Earth coalesced nearly 5 billion years ago, the early universe existed only as ions. For example, oxygen existed as individually charged particles, not as atoms or molecules. As oxygen cooled into its most chemically stable atomic form, it bonded rapidly to atoms of other elements to form oxides, which were incorporated into Earth's crust. Because many elements form their most stable compounds with oxygen atoms, oxygen soon became the most abundant element in the crust, accounting for virtually half its weight. In the atmosphere, atomic oxygen bonded with hydrogen to form water and with carbon to form carbon monoxide and carbon dioxide. Only a few free oxygen molecules remained available in the atmosphere for the energy transfer reactions that present-day microorganisms would need.

Within less than 1 billion years of Earth's beginning, the first living cells formed. They were soon followed by a rich variety of one-celled archaea and bacteria. By the time Earth was 2.5 billion years old, light-loving bacteria that used molecules like chlorophyll to harvest energy from the sun had evolved into cyanobacteria (formerly classified as blue-green algae). Using light for energy and an iron-sulfur protein, cyanobacteria harnessed the chemical reaction that changes water and carbon dioxide into organic matter (carbohydrates) for soil and into molecular oxygen (O_2) for the atmosphere. Other organisms fed, in turn, on the carbohydrates. However, the cyanobacteria would need more than a billion years to produce enough molecular oxygen

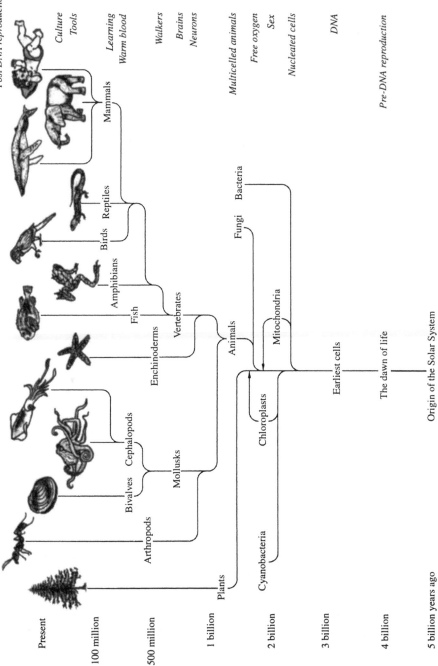

Post DNA reproduction

Culture
Tools

Learning
Warm blood

Walkers
Brains
Neurons

Multicelled animals

Free oxygen
Sex

Nucleated cells

DNA

Pre-DNA reproduction

Mammals

Reptiles

Birds

Amphibians

Fish

Enchinoderms

Vertebrates

Bacteria

Fungi

Mitochondria

Animals

Cephalopods

Bivalves

Mollusks

Chloroplasts

Arthropods

Earliest cells

The dawn of life

Cyanobacteria

Plants

Origin of the Solar System

Present

100 million

500 million

1 billion

2 billion

3 billion

4 billion

5 billion years ago

to accumulate in the atmosphere and to initiate the development of life forms we know today. It was iron that caused the long delay. The early oceans contained so much ferrous iron (Fe^{++}) that the cyanobacteria needed 750 million years to produce enough oxygen to precipitate it all out in its ferric (Fe^{+++}) form (see figures I.1 and 2.1).

The Oceans Precipitate Iron

As long as the atmosphere lacked oxygen, the iron in the oceans and other bodies of water remained in its ferrous state (Fe^{++}). However, as photosynthetic bacteria began to produce molecules of oxygen, the oceans' ferrous iron started turning to rust, which is so extremely insoluble in water that it precipitated out. The globe still bears the marks of this colossal rusting process.

About 2.7 billion years ago, precipitating ferric oxides began to build the intensely red deposits called banded iron formations. These rich sources of iron ore are located around the planet: near the Great Lakes of North America and in Labrador, western Australia, and Mauritania in northwest Africa. These deposits date the rise of the cyanobacteria and the oxygenation of the atmosphere. Later, ferric oxides also formed the red soils and rock formations of the southwestern United States.

Eventually, about 2 billion years ago, oxygen exhausted the oceans' supply of ferrous iron. At that point precipitation ceased, leaving only about one thousand free iron atoms in each cubic centimeter of seawater. The molecular oxygen produced by cyanobacteria could finally begin to accumulate in the atmosphere. Oxygen quickly reached its current level of 21 percent between 0.5 and 1.5 billion years ago. Thus, the oxygen atmosphere that we rely on today is a relative newcomer: it has been Earth's companion for only about a billion years, less than one-quarter of Earth's life (see figure I.1).

FIGURE 2.1. THE EVOLUTION OF LIFE ON EARTH IN RELATION TO THE ORIGIN OF THE SOLAR SYSTEM.
Adapted from Hans Moravec, Mind Children *(Cambridge, Mass.: Harvard University Press, 1988). Copyright © by Hans Moravec. Reprinted by permission.*

Oxygen Molecules Flood the Atmosphere

As oxygen flooded our atmosphere, anaerobic archaea and bacteria faced a life-or-death crisis. Oxygen is highly toxic to biological molecules. Inside living cells, it can produce free radicals, atoms or molecules with loose extra electrons. Armed with their extra electrons, free radicals can damage essential cellular components such as proteins, lipids, and nucleic acids. Free radicals have been blamed for causing a wide range of problems, from aging to cancers. To survive the onslaught of oxygen in Earth's atmosphere, anaerobic organisms faced two choices: they could hide from the oxygen, for example, in mud and rocks or beneath the seafloor. Or they could evolve new metabolisms by switching to an oxygen-based biochemistry. Many—probably most—of the early world's organisms were unable to do either and failed to survive.

Considering how difficult it is for species to radically alter their metabolisms, the advantage gained by doing so would have to be great. Why did oxygen-based life win the evolutionary race on Earth? Oxygen won because it is a powerhouse of energy. Organic matter, that is, carbohydrates or glucose, can produce eighteen times more energy in the presence of oxygen than in its absence. Thus, when molecular oxygen oxidizes something else, it invariably gives off energy. The transition from anaerobic to oxidizing aerobic life was an enormous evolutionary step on Earth. It opened up vast reservoirs of oxidizing energy for living organisms. It made the rich diversity of Earth's flora and fauna possible. Iron played a vital role in the process. The molecule responsible for 90 percent of the cell's molecular oxygen uptake is cytochrome oxidase, a molecule which has a center of iron and copper. This molecule binds molecular oxygen so tightly between its copper and iron atoms that the cell is protected from toxic free oxygen (O).

The increased energy efficiency of the planet's new oxygen-rich atmosphere came at an enormous price, however. Organisms still needed iron. They could not abandon 3 billion years of iron-based biochemistry outright. But the iron that had been readily available in its ferrous form on anaerobic Earth was now rust and chemically unavailable. And so much ferric iron had precipitated out of the ocean that the only chemically available iron had become a trace element in

RESPIRATION

Respiration is the consumption of a food source in the presence of air. More specifically, it is the reaction of molecular oxygen (O_2) with carbon compounds to form carbon dioxide and water. The reaction releases large amounts of energy, just as burning charcoal briquettes on an outdoor grill gives off enough heat to cook food. Respiration enables cells to use carbon, generally from glucose and fructose, to manufacture amino acids and nucleotides, the building blocks of proteins and nucleic acids, respectively. DNA and RNA, which use amino acids to create proteins, are important organic compounds that contain a phosphoric acid group. In short, respiration constitutes a veritable banquet and a fabulous source of energy for cells.

the biosphere. Except in the sea-vent communities, there was far too little iron to be of practical use to living cells.

Paradoxically, oxygen gave our watery planet a biochemistry that is both oxygen based and iron dependent. We became an evolutionary curiosity, a planet that forms rust at global levels. Despite a wealth of iron, however, little of it was biologically available. Thus, when we turn in the next chapter to the iron-laden surfaces of huge geological formations and to the molecular world of microorganisms, the most precious metal on Earth will still be iron. The need for iron will be so pervasive that competition for the metal will become fiercely competitive. Creatures that had evolved a primitive, iron-based biochemistry will be forced to build special molecules, weapons to control Earth's newly scarce supplies of iron. As iron's journey through the biological world continues, the arrival of molecular oxygen in the atmosphere inaugurates life's battle to control its iron supplies.

3

Grabbing and Storing
CONTROLLING IRON

Virtually all living organisms walk a tightrope between iron starvation and iron toxicity. Too little iron, and cells starve. Too much iron, and cells are poisoned. To survive, life forms ranging from humans to the smallest microorganisms must strictly regulate their iron supplies. Moreover, they must do so in an oxygen-rich atmosphere in which free iron is extremely scarce. It is no wonder that ancient organisms evolved three special molecular marvels—siderophores, ferritins, and transferrins—for seizing, storing, and recycling iron atoms.

The vast scale of these iron-seizing and -regulating systems is quite startling. Each day the typical healthy human manufactures 200 billion red blood cells, each one of which contains many trillions of iron atoms. Some bacteria accumulate 1 percent of their weight as iron. Yet thanks to the arrival of oxygen in our atmosphere, assimilating iron from the natural environment can be extremely difficult. As a result, all forms of life are plagued by iron deficiency. Even plants can get iron "anemia," which turns their leaves yellow. Iron deficiency anemia is a common nutritional problem among the poor in the developing world. And the molecules that seize, store, and transport iron in living systems play a role in a large number of human diseases, including rusty liver disease, hemochromatosis, diphtheria, meningitis, bacteremia, enterocolitis, and the acute diarrheas caused by *Escherichia coli* and *Campylobacter jejuni* bacteria.

Iron Grabbers: Siderophores

In the relentless search for iron, microorganisms must wage war against each other to grab one another's iron supplies. When iron levels fall too low, most microbes produce iron-grabbing weapons and battle one another for iron. Their weapons are called siderophores. Because siderophores are made by almost all microorganisms, they pervade the living world, including the bodies of human beings.

J. B. Neilands discovered siderophores when he was a young instructor at the University of Wisconsin in 1951. Neilands had previously worked in Sweden with an authority on the important iron-containing protein cytochrome c, which plays a role in energy-producing reactions inside cells. With the help of cytochrome c, iron atoms cycle their electrons and change back and forth between the reduced (Fe^{++}) and oxidized (Fe^{+++}) states to energize chemical reactions in living organisms.

Neilands has a dry sense of humor, and as he puts it, he achieved "some notoriety" by discovering the method that is still used today to purify and isolate cytochrome c. Soon after his discovery, he moved from Sweden to the University of Wisconsin, where he joined the faculty as a lowly instructor. Because of his experience with cytochrome c, his new boss told him to help a researcher in the botany department. Neilands recognized an order when he heard one and hastened upstairs to visit Paul Allen. Allen was studying a fungus, *Ustilago sphaerogena,* that infects grasses. Under certain conditions, his fungus culture kept turning pink. Allen wondered if it contained cytochrome c. Assuming that cytochrome c would grow only in some exotic media, Allen thought he needed Neilands's expert help. Neilands, however, promptly dumped the fungus into a generic all-purpose formula, and sure enough, the fungus culture turned baby pink. "There was nothing magic about the media; you could grow the fungus on anything," Neilands said. But after isolating the iron protein, Neilands realized that it was cytochrome c all right. Nevertheless a lot of red color remained, and Neilands wondered what it was.

Next Neilands grew the fungus in the presence of iron. The fungus produced outrageous amounts of cytochrome c—up to 1 percent of its own weight. To compete with the fungus, the body of a 200-pound person would need to incorporate 2 pounds of iron, far more than a human's normal 3 to 4 grams (about one-tenth of an ounce). This was no ordinary fungus. Furthermore, when starved for iron, it excreted large amounts of a new colorless substance. And when iron was added back into the media, the mysterious substance turned red again. Obviously the mysterious red substance was associated with iron.

By throwing the substance into solvents and then drying and extracting it, Neilands was able to produce crystals of the first known siderophore. (In 1973, scientist Charles E. Lankford coined the term siderophore from the Greek for "iron bearer.") Neilands's siderophore can bind atoms of ferric iron (Fe^{+++}) more tightly than almost any known cellular substance. Iron-starved fungi and bacteria manufacture siderophores to grab hold of any available iron. Neilands's siderophore was wildly effective. "It was serendipitously and fortuitously isolated. I wasn't looking for it. I just came across it in this fungus," Neilands said modestly. It has taken thirty years of coordinated research by biochemists, microbiologists, and physicists to demonstrate that almost all microorganisms manufacture iron-grabbing siderophores. In fact, some produce staggering amounts, and many microbes make three different siderophores, probably as a backup system. In addition, evolution has made microorganisms expert at identifying and absorbing their own particular siderophores. If the slightest detail in the siderophore's structure is changed, its microorganism may reject it outright.

Siderophores have one of the most difficult chemical jobs on our planet. Anyone who has tried to remove rust penetrating a piece of metal can appreciate that. Siderophores can break down rust (ferric oxide); they can grab iron from oxides in soil, liquid, or living tissue. Some siderophores do the job better than the strongest commercially available rust dissolver, EDTA (ethylenediaminetetraacetic acid). EDTA softens water, removes unwanted metals, and is used to treat people with metal poisoning.

The notorious intestinal bacteria *E. coli* makes a siderophore that can grab iron trillions of times more powerfully than EDTA. *E. coli* lives in the intestine, where the only available iron occurs in digested food. Its main siderophore, enterobactin, can capture iron from digested food. Enterobactin is the world's most potent siderophore. Some *E. coli* strains—such as those that cause meningitis, bacteremia, and other human infections—are especially virulent because they have three kinds of siderophores. Why does *E. coli* grab iron so desperately? Perhaps because it competes with its host for iron.

Much of our knowledge about the iron in bacteria and the human body comes from experiments conducted with *E. coli*. When microorganisms find themselves in a low-iron environment, their cellular

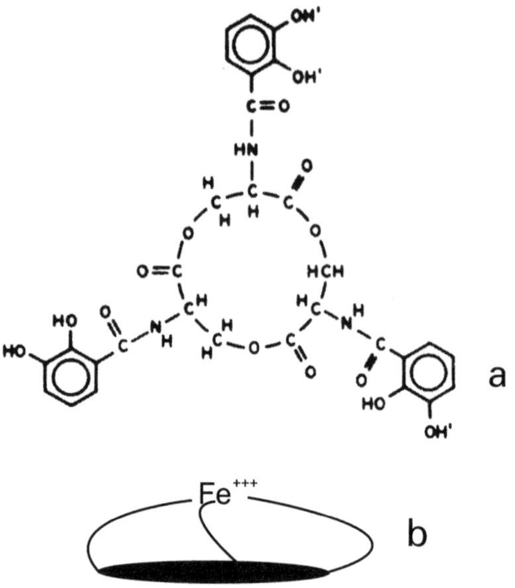

FIGURE 3.1. ENTEROBACTIN, *ESCHERICHIA COLI'S* SIDEROPHORE:
(a) E. coli*'s siderophore before it grabs the iron;* (b) E. coli*'s siderophore grabs the iron. The black base in* b *represents the central core of the siderophore; the central ring in* a *is rather flat until it raises its three branches up from its central ring to grab an atom of ferric iron.*

chemistry sends a signal. The organism responds by manufacturing a siderophore. These large molecular iron-grabbers are flat when empty, but when a molecule containing iron drifts near, the siderophore raises three arms and snatches the iron out of the molecule with a spectacular vise-like grip (see figure 3.1). With these three upraised arms, the siderophore mimics rust's molecular hold on iron. The siderophore bond is so strong that the siderophore can even pull iron out of rust.

While a siderophore's iron-seizing job is difficult, delivering the captured iron safely to the cell is also challenging. For decades, no one understood how an iron-laden siderophore can get inside the cell of a bacterium. A series of membranes allows the cell of a bacterium to feed from its external environment while protecting its internal chemical factory from outside toxins such as antibiotics, disinfectants, and detergents.

Fat with iron, siderophores must squeeze their way into bacterial

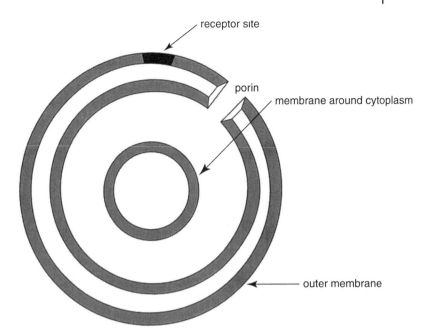

FIGURE 3.2. A BACTERIAL CELL.
This schematic shows the organization of some important parts of a bacterial cell that are discussed in this book.

cells through skinny gates, called porins, that are only a tenth as wide as a siderophore. Then, in 1997, Philip Klebba, a biochemist at the University of Oklahoma, watched as an iron-loaded siderophore slithered into an *E. coli* cell. The membranes of bacterial cells often have gates that open and close in response to particular chemicals. These gated channels stick out from the cell's surface like long, tubular mouths. A small floppy molecule, a bit like a soft cork or flexible hinge, plugs the mouth shut. When the plug changes shape, the channel opens for roughly five minutes, long enough for an estimated 50,000 iron-laden siderophores to slip inside the cell. Once the siderophore is inside the porin, a special protein called ton B helps the siderophore enter the innermost part of the cell, the cytoplasm. Somewhere on the journey between the outer membrane and the cytoplasm, the iron separates from the siderophore and is incorporated into other iron complexes (see figures 3.2 and 3.3).

Cell exterior

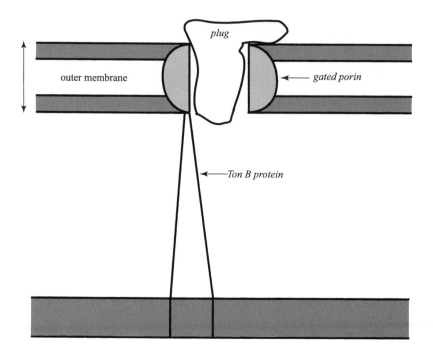

Cell interior

FIGURE 3.3. A GATED PORIN CHANNEL.
At a chemical signal, the plug changes shape, and the gated channel allows a siderophore (not shown) to pass through the cell's membranes. A ton B protein extends from the membrane surrounding the cell's cytoplasm to help the siderophore enter the cytoplasm. Adapted from H. Nikaido and M. H. Saier Jr., "Transport Proteins in Bacteria," Science 258 (1992): 936–942. Copyright 1992 by the American Association for the Advancement of Science. Reprinted with permission.

Armed with their siderophores, bacteria can capture the iron they need to survive on our oxygen-rich planet. The siderophores help bacteria balance on the narrow tightrope that divides iron starvation from iron overload. But what happens if an organism absorbs too much iron?

In 1962, medical reports about Bantu tribes in South Africa provided the first indication that too much iron in the human diet can be toxic. The men of these tribes drank large quantities of homemade beer brewed in iron pots. Because fermentation had leached iron from the pots into the beer, the men ingested so much iron oxide that their livers eventually looked rusty. According to many medical observers, the livers had "rusted away." Rusty liver disease has also been reported in other countries. Recent indications suggest that many sub-Saharan Africans may have a genetic predisposition for absorbing abnormal amounts of iron.

Iron Storers: Ferritins

Excess iron is extremely dangerous because it destroys cells. What protects living organisms from iron poisoning? A cell's strongest defense against the toxicity of excess iron is a special class of iron-storage molecules called ferritins. Ferritins are made of a huge, hollow protein shell, a veritable rent-a-storage container of a molecule. Ferritin molecules are so large that they are comparable in scale to many cells and are almost as large as a virus (see figure 3.4). A human ferritin, which measures twelve-billionths of a meter in diameter, can easily cache away two thousand iron atoms. Horse ferritin is especially commodious: it holds about five thousand iron atoms. In

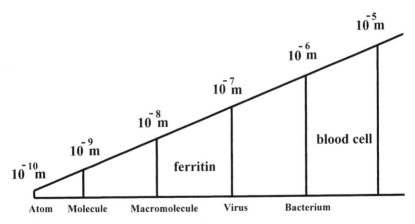

FIGURE 3.4. A BIOLOGICAL SCALE.
Ferritin is a large molecule, almost as big as a virus.

FERRITIN, SUPERPARAMAGNETISM, AND INFORMATION STORAGE

Lining up like toy soldiers in tiny iron hydroxide microcrystals, the iron atoms inside the core of ferritin form an unusual magnetic system. If enough of these ferrihydrite crystals cluster together, they form a weak magnet called a superparamagnet. These microcrystals are far too small to see in a light microscope. At room temperature the magnetic alignment of these molecules is easily disrupted by thermal energy. But place the ferritin molecule in a magnetic field, eliminate thermal disruption by lowering the temperature, and the microcrystals magnetically line up. Placed in a magnet, their electron spins can swivel freely, pointing in the same direction as the prevailing magnetic field.

Theoretically, provided the temperature was low enough, a film of ferritin molecules could store information as precisely as magnetic recording tape. Lines or regions of ferritin molecules could be magnetized so that all their electron spins pointed in one direction. Elsewhere on the "tape," the spins could be aligned in the opposite direction. Thus information could be encoded and read. Bacteria raised in iron-rich soup could manufacture ferritin magnetic tape molecules the way conventional factories make magnetic tape.

The existence of these magnetic domains was predicted decades ago by two solid-state physicists, Charles P. Bean and Israel S. Jacobs, who were studying supersmall magnetic particles. They predicted, correctly, that any sphere of magnetic molecules roughly 100 angstroms across would form a minimally sized magnetic domain. Bean and Jacobs came extraordinarily close. Ferritin measures approximately 120 angstroms across. But Bean and Jacobs could never have imagined in 1963 that magnetic domains would be discovered in these iron-storage molecules or that virtually every living cell would use them.

mammals, ferritin stores all this iron away, amazingly enough, as a rather inert ball of rust. When iron enters a ferritin closet, it is oxidized from ferrous to ferric iron. Interestingly, inside ferritin, the iron also forms an unusual magnetic system, called superparamagnetism, which is different from the more usual kind of ferromagnetism found in iron ores and lodestone. When the organism needs more iron, the closet allows the metal to escape for recycling. Thus, as long as a cell is armed with ferritin, it has both an iron-storage and an iron-resupply system.

Ferritin storage molecules are virtually universal among life forms, and their ubiquity provides a clue to their importance and their ancient origins. Ferritin molecules are found in bacteria, fungi, plants, invertebrates, and vertebrates—everywhere but in lactobacilli, milk bacteria. Bacteria keep an especially tight rein on their iron metabolism; the intestinal bacteria *E. coli* not only has three kinds of siderophores to capture iron but also has two kinds of ferritin to store it. Because vertebrates do not make siderophores, they rely heavily on the simple elegance of ferritin's storage and retrieval system. In mammals, ferritin stores about 15 percent of the body's iron and nearly 90 percent of the iron released when red blood cells die.

Humans have no biochemical system for transforming iron into a benign substance for excretion. For example, patients undergoing medical treatment sometimes produce reddish urine and black stools; these are clues that excess iron was excreted chemically unaltered. A biochemical excretion process would be far more complicated than ferritin's storage system. Because iron is extremely insoluble in water, the average person would have to drink 2.63 trillion gallons of water daily to excrete enough iron to prevent toxic overload. Given these difficulties, it is not surprising that vertebrates failed to develop a biochemical system to get rid of excess iron.

Thanks to their practicality and efficiency, ferritin molecules function in much the same way wherever they are, whether in bacteria, plants, invertebrates, or vertebrates. The elegant simplicity of ferritins has attracted a band of hardy admirers. The English crystallographer Pauline Harrison, who has spent five decades studying the structure and dynamics of the ferritin molecule, is known—unofficially, of course—as "The Iron Lady." Harrison's work with ferritin began at

Oxford University under the supervision of the late Dorothy Crowfoot Hodgkin, who won a Nobel Prize in chemistry for deciphering the molecular structure of several medically important substances: penicillin, insulin, and vitamin B_{12}, the cure for pernicious anemia.

In 1949, the father of ferritin biochemistry, Sam Granick of Rockefeller Institute in New York City, gave Hodgkin some large ferritin crystals grown from horse spleen. Back home, Hodgkin handed the crystals to Harrison and told her to figure out as much about its molecular structure as possible. The job would not be easy though. No protein structure had yet been determined, and techniques, such as x-ray crystallography, for determining the structures of large molecules were in their infancy. Harrison did not know whether the job was even possible.

She explained, "They were very nice big crystals but x-ray machines were very low intensity in those days, so it took about 24 hours to get the kind of photograph you'd now get in a minute or two at a synchrotron." Harrison was still working in Hodgkin's laboratory when John Farrant, an electron microscopist in Melbourne, Australia, discovered that ferritin is essentially a large ball of "rust" coated with protein.

Harrison was fascinated. "Ferritin has a very beautiful structure with high symmetry so it's very pleasing as a molecule to look at." She thought of it as a big bag or as "a big, nice, round, shiny chestnut, an iron core with an outer protein shell that comes apart in pieces and has holes between the subunits where we think iron may go in and out."

For some years, Harrison frequented slaughtering houses for ferritin-rich horse spleens to chop up. As she recalled, "You got a nasty smelly mess, especially if you got old horses because they were full of the stuff." During the late 1980s, researchers in her lab "cloned" human ferritin, a feat that provided a simpler way of obtaining the protein and opened experimental avenues for understanding the iron pathways in human biochemistry.

Harrison and her colleagues discovered that the iron atoms move in and out of ferritin through fourteen channels. Like woolen clothing being put away for the winter, the atoms move through these corridors for storage in the molecule's inner closet. This easy ebb and

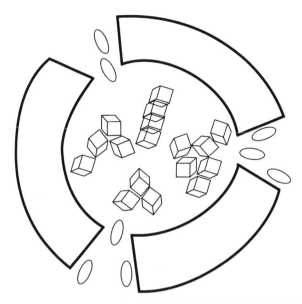

FIGURE 3.5. A SCHEMATIC, CROSS-SECTIONAL VIEW OF A FERRITIN IRON-STORAGE MOLECULE, SHOWING THREE OF THE MOLECULE'S FOURTEEN ENTRANCE CHANNELS.

Protein subunits surround a hollow shell that has an iron microcrystal forming inside it. Human ferritin, which measures 12 nanometers, or 12 trillionths of a meter in diameter, can hold about 2,000 iron atoms. The ferritin in horse spleen accommodates about 5,000 iron atoms. The ovals represent iron atoms entering the ferritin. The cubes represent subsequent crystallization of this iron. Adapted from P. M. Harrison and T. H. Lilley, "Ferritin," in Iron Carriers and Iron Proteins, *ed. T. M. Loehr, 125–238 (New York: VCH, 1989). Copyright © John Wiley and Sons, Inc. Reprinted with permission.*

flow through channels keeps the cell's iron supply safely in balance (see figure 3.5).

Iron Transporters: Transferrins

Our understanding of iron's function within the human body expanded with breathtaking speed during the 1990s as scientists studied iron metabolism from a molecular point of view. Molecular and cell biology and particularly recombinant-DNA technology revealed how living cells metabolize iron, how genes regulate the process, and how aberrations in the system can cause disease. Another class of molecular iron-transporters, transferrins, emerged as key players in the process.

Transferrins operate a shuttle service to recycle iron atoms back and forth between their ferritin cages and other parts of the body. Their pun-like name, "transferrin," explains their iron-transferring function quite nicely. Transferrin molecules are organized into two very similar lobes of amino acids. Each lobe contains an iron-binding site, so that each transferrin molecule can carry two iron atoms. The process of iron transport begins when food enters the uppermost part of the small intestine, the duodenum. There, gastric acid helps incorporate the iron from inorganic compounds into compounds that are more soluble in the alkaline environment of the duodenum. The small intestine is lined with roughly 5 million small, hair-like projections called villi. The villi are tipped with mucosal cells that have a protective lubricant to help absorb iron and other nutrients. Because the villi are replaced every five or six days, iron supplements are best taken weekly rather than daily.

As iron atoms cross the cellular lining of the intestinal wall, they are captured by transferrin molecules and delivered to the plasma, the clear fluid component of blood, and from there either to the liver cells where they are stored in ferritin or to the bone marrow where they are incorporated in red blood cells. When a red blood cell dies after its allotted life span of about 120 days, it also releases iron into the bloodstream. Transferrin picks up this iron too and recycles it in the same way. Cells—for example, liver cells and developing red blood cells—assimilate iron from transferrin at special receptor sites studding the cells' surface, in a process called endocytosis (see figure 3.6). If a cell is iron deficient, an *iron-sulfur protein* activates an *iron-regulatory protein* that increases the number of transferrin receptors on the cells' surface. When iron is plentiful, this same regulatory protein blocks the production of transferrin receptors and increases the number of ferritin storage molecules. This reciprocal control over iron levels in cells is exquisitely micromanaged by an iron-sulfur molecule of extremely ancient lineage. Cells that need large amounts of

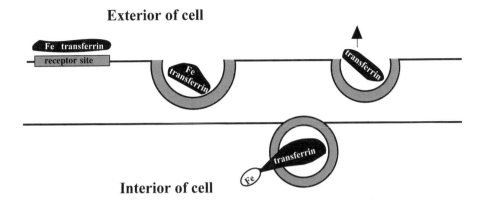

FIGURE 3.6. ENDOCYTOSIS.

A cell extracts iron from transferrin by means of endocytosis and later releases it. During endocytosis, the cell swallows the receptor site, its transferrin, and its iron in one giant bite by stretching its membrane around the entire unit, scooping it up, and shoving it inside. Within minutes, the iron is plucked from its transferrin and receptor site. Left holding the empty transferrin and receptor site, the cell pushes them back to the surface, and the receptor site releases the transferrin, which heads back into the bloodstream to be reloaded with iron. Mammalian cells also use endocytosis to vacuum up large foreign particles, large microorganisms, and dead cells.

iron—for example, developing red blood cells in the bone marrow—produce a particularly large number of receptor sites.

At any one time, the plasma holds about 3 milligrams of iron, but 30 milligrams of iron in transferrin molecules can race through this small iron pool every day. Generally, only about a third of transferrin molecules are saturated with iron. Theoretically, transferrins could hold much more iron, but if they are more than 55 percent saturated, a person may be at risk for toxic free iron. Free iron—iron not stored in ferritin, transferrin, hemoglobin, or other molecules—catalyzes several cellular reactions that form dangerous free radicals. Free radicals can damage cellular components by stealing electrons from other atoms and molecules. In living systems the most convenient atoms to steal electrons from are hydrogen atoms, many of which are in DNA strands. When an electron is stolen, the DNA strands get nicked and cell membranes are disrupted.

Transferrin molecules come in several varieties, depending on whether they transport iron in bird egg white, human tumors, blood serum, or elsewhere. Lactoferrin, for example, is found in milk, tears, saliva, pancreatic juices, and mucus. Lactoferrin binds iron atoms especially tightly and may thwart the growth of many disease-causing bacteria, and help infants fight bacterial infections by reducing the amount of available iron.

Iron in the Balance

When iron supplies are in balance, a healthy person has enough iron for metabolic activities and the immune system but not enough to encourage the growth of pathogenic microorganisms or liver tumor cells. The bodies of healthy people continually adjust the number of ferritin and transferrin molecules in their blood.

Healthy people who are mildly anemic make fewer ferritin storage molecules and more transferrins than normal. The liver makes more transferrin to scavenge more free iron from the blood plasma, and the number of iron-receptor sites is maximized. In contrast, healthy people with slightly too much iron produce more ferritin storage molecules; properly regulated bodies never lack storage space for iron.

The concentration of ferritin storage molecules in the blood serum provides a test for iron deficiency or overload. Extremely high levels of ferritin in blood serum may indicate dangerous conditions that have overwhelmed the storage capacity of ferritin molecules. And even adult males and postmenopausal females who appear healthy should be checked for iron deficiency anemia because it is often the first sign of gastrointestinal bleeding, which can be caused by gastrointestinal cancers. (In industrialized nations, gastrointestinal bleeding is the leading cause of iron deficiency in men and the second leading cause in women.) Thanks to the telltale anemia these cancers cause, they can often be diagnosed while the tumors are still operable.

Bacterial conditions can upset the body's best-laid plans for keeping its iron stores in balance. Preying on the body's iron supply system, for example, some invading pathogens have developed iron grabbing into a martial art. The virulence of diphtheria, a childhood disease that before immunization was often fatal, increases in iron-depleted environments. Some pathogens nab siderophores and iron-binding proteins from other microorganisms. Enterocolitis and the painful diarrhea and fever of *Campylobacter jejuni* are caused by bacteria that steal siderophores from other bacteria. Some bacterial species in blood plasma hijack the iron straight out of nearby transferrin and ferritin molecules, bypassing siderophores altogether. Some bacterial siderophores are so powerful that they can nab iron even from Desferal, an iron-chelating drug.

Is the person with abnormally high or low iron levels more susceptible to infectious and inflammatory disease? Surprisingly, iron's effect on the immune system, infection, and inflammation is widely debated by physicians and biochemists. Iron-overload diseases do not seem to increase a person's susceptibility to infections. And lowering an individual's ability to capture iron may not reduce infection either. Studies conducted in the 1970s seemed to show that a high fever could limit the body's ability to absorb iron. Scientists wondered if this could be the body's natural method for controlling the growth of pathogenic organisms. However, the possibility of convulsions and brain damage was ignored, and few comprehensive or conclusive studies were conducted.

Genetic conditions can also cause iron imbalances. Hereditary hemochromatosis, the most common genetic disorder in the United States, overloads approximately 1 million Americans with excess iron. Hereditary hemochromatosis is triggered by a defective gene that causes one amino acid to be substituted for another in a protein. Many Hispanics and one in five people of northern European stock carry the mutated gene, and one in two hundred or three hundred Americans inherits the gene from both parents and thus suffers from the disease. Most of them inherited the defective gene from a single common ancestor who lived centuries ago.

Hemochromatosis makes the intestines absorb twice as much iron from food as normal. Early symptoms—fatigue, sore joints, and frequent infections—are ephemeral and easy to miss. But as the iron builds up in the liver, heart, spleen, and pancreas, it destroys cells. In severe cases, a patient may have two hundred times too much iron before the condition is identified. By the time a person is fifty or sixty years old, the organs may have literally rusted. Untreated, hemochromatosis can cause cirrhosis of the liver, liver cancer, heart failure, diabetes, arthritis, and eventually death. Liver and pancreas cells do not regenerate, so any damage is permanent. The yellowish skin and diabetes-like symptoms that characterize the disease have given it the nickname bronze diabetes.

Although carriers of the gene can transmit it to their children, they generally do not show any symptoms of hemochromatosis. Some carriers do, however, have slightly more iron in their blood than normal. Inheriting one hemochromatosis gene has probably helped millions of women with iron-poor diets survive menstruation, pregnancy, childbirth, and lactation. Because of menstruation, women who inherit two genes may not develop symptoms of hemochromatosis until after menopause.

In contrast, men who have undiagnosed and untreated hemochromatosis may suffer organ damage in their fifties. Screening all thirty-year-old white males (twenty years before their iron levels become toxic) would save society between ten thousand and forty thousand dollars for each life spared. The blood test for hemochromatosis measures the amount of iron in the transferrin and the amount of ferritin in the body. As a result of the recent identification of the gene for hemochromatosis, a genetic test is being developed.

The treatment for hemochromatosis is bloodletting, a simple, straightforward procedure that has long been viewed as a barbarous custom of ignorant times past. But bloodletting, technically called phlebotomy, does bring the iron levels of hemochromatosis sufferers down to normal. One blood unit (500 cubic centimeters of blood) is removed once or twice weekly for about one year. Thereafter, donating one unit of blood every other month maintains optimum iron levels. Thanks to bloodletting, hemochromatosis is a controllable disease, provided it is diagnosed early enough.

Until a reliable and simple genetic test for hemochromatosis becomes available, it is important to monitor the iron status of men after the age of thirty or forty and of women after the age of forty or fifty. Products marketed as "vitamins for men" are generally iron free. Some physicians also recommend that vitamin C supplements, which enhance iron absorption, be given only to anemic patients.

Siderophore iron-grabbers, ferritin storage units, and transferrin transporters allowed bacterial cells to adapt to the new, oxygen-rich atmosphere on Earth. With these molecular weapons, cells could capture scarce iron and hold it ready for use when needed. They could maintain a balance between too little and too much of this essential, toxic metal. They could also tightly control their host organism's iron supplies, for any disturbance to this exquisitely balanced system could precipitate serious problems.

Some creatures, however, were more timid. They wanted to control their iron supplies and, at the same time, limit their exposure to atmospheric oxygen. So here our story of iron's journey through the biological world must move to a more sophisticated level of biochemistry and focus for the first time on an entire, albeit primitive, organism. We will look at living creatures so eager to control their environment that they learn to build complex iron structures composed of many molecules. These are the magnetic bacteria, some of nature's charmers, tiny creatures that manufacture iron magnets in order to stick in the mud and the oxygen-poor waters that they love.

4

The Smallest Living Magnets
AVOIDING OXYGEN

In the battle against oxygen, some microorganisms went to enormous lengths to escape the onslaught of the powerful new molecules. So desperate were they to limit their exposure to atmospheric oxygen that they learned to manufacture miniature magnets.

"I see it crystal clearly," Richard Blakemore said, recalling the evening he discovered Earth's smallest living magnets. "I get excited every time I look at them."

It was already dark outside the laboratory as Blakemore, peering through his microscope, searched through mud samples for bacteria. At twenty-three, Blakemore was a second-year graduate student in microbiology at the University of Massachusetts in Amherst. In 1975, fledgling microbiologists there were often assigned such simple tasks as identifying the material between their teeth or analyzing organisms in mud. His professor had collected the mud from a Massachusetts marsh and asked Blakemore to learn everything possible about some large spiral bacteria in it. But that night, Blakemore said, "other organisms forced their existence on me."

Microbiologists study living organisms of microscopic size: bacteria, protozoa, viruses, and microscopic algae and fungi. Bacteria—more formally called procaryotes—are primitive cells without nuclei. They, or possibly another group of ancient life called archaea, were the first life to appear on Earth 3 or 4 billion years ago.

Several days into his project, Blakemore had become "sort of aware of these little one-celled organisms that were swimming around, that were my most problematic contaminant. . . . They appeared to be most numerous. I was working at low magnifications—gradually—it kind of dawned on me that these organisms were more or less moving in a unidirectional way." Each time he put a new drop of muddy water under his microscope, a few "little points of light" appeared. But soon hundreds and even thousands of them would come streaming across the slide, always rushing in one direction—strange contaminant indeed.

Working days, Blakemore had assumed the organisms were responding to the light that shone through the tall windows of the laboratory. Bacteria, among the simplest of organisms, were known to move in response to light and to higher or lower concentrations of various chemicals. But that night, the windows were dark. So the "little points of light" could not possibly be racing toward the windows.

To check, he tried covering the microscope with a pasteboard box and shielding it with his hands, but they continued to move in one direction. "What were these things doing in here? Why are they always moving to one side? That's really strange," he thought. Although he kidded an undergraduate working in the lab about "north-swimming bacteria," he was still sure the organisms responded in some way to light. Eventually, Blakemore called over the third student working in the lab that night and asked, "How would you explain that?"

With the majestic authority of an advanced graduate student, John Bresnick ordered the tyros, "Away from the scope." He suspected that they were tilting the microscope, letting the water slide and the organisms stream to one side. When Blakemore repeated his joke about "north-swimming bacteria," the undergraduate countered, "Well, if they're north-swimming, we should be able to pull them with a magnet, right?"

So while Blakemore looked through the scope, the student picked up a magnetic stirrer lying beside the microscope and brought it up behind the swimmers. "Fortunately," Blakemore recalled, "he had the end of the magnet pointing toward them so that it attracted them. And—all of a sudden—en masse—this whole massive population of bacteria swims in exactly the opposite way across the microscope stage. It was incredible, just incredible, and no one even believed my response. They thought I was kidding—until they looked in." Then they too saw light specks race to one side of the water drop and, when the magnet was reversed, instantaneously turn and race to the other side of the drop. Back and forth they rushed like schools of tiny, silvery fish, back and forth with the magnet. At that point, John Bresnick said, "I think you've discovered something."

No one had ever dreamed that bacteria could be magnetic. During the 1960s, it had been reported that a number of creatures—including a protozoan, a mud snail, a scarab beetle, and fruit flies—were sensitive to weak magnetic fields. In addition, particles of the magnetic iron oxide magnetite had been discovered in the teeth of the world's largest chiton, *Cryptochiton stelleri,* a 20-inch-long mollusk that lives in intertidal zones of the northwest Pacific. But the chiton's teeth had been dismissed as an isolated curiosity. The idea that living organisms—much less lowly bacteria—might make and use compasses seemed preposterous. Yet Blakemore had indeed discovered magnetic bacteria.

"From then on," Blakemore said, still starry-eyed more than twenty years later, "it was a night of incredility."

Technically, Blakemore should not have been able to see the bacteria at all. His optical, dissecting microscope magnified them less than 100 times. But he was looking at them that night against a dark background, and the light of his microscope bounced off their cell walls. Almost every bacterium has a rigid cell wall surrounding and protecting the plasma membrane that contains the cell's genetic material. Bound into this plasma membrane are flagella, whip-like arms that enable the bacteria to move. Thanks to the low magnification, Blakemore could spot light reflected off their cell walls. As he realized later, "In order to see them, you won't want to use a high magnification. You really get a much more realistic appreciation of them all doing the same thing as a population of cells if you look under low magnification." With a more powerful microscope, he would have seen more details and would not have realized that masses of the bacteria were moving as a school in one direction.

Blakemore's microbiology professor was in Italy at the time, and Blakemore was exploding with the news, so he raced home to tell his wife Nancy. Abandoning all grammar in the joy of the memory, Blakemore said, "It couldn't have been perfecter. I didn't really—hardly—know how to take it in."

Racing through books on bacteria, Blakemore read that they could respond to electricity, light, gravity, and temperature, but he could find nothing about magnetism.

"I wasn't sure they were bacteria either," Blakemore realized. To get better samples for study, he began separating the magnetic bacteria from all the other living organisms in the mud. When he brought

GROWING MAGNETIC BACTERIA

Take a one-and-a-half quart pickle jar to a muddy marsh, preferably one with black soil. Fill the jar more than half full with mud. Then cover the mud with several inches of water from the same area. Do not fill the jar completely; leave some space at the top. Put the lid on and place the jar on a shelf at room temperature for at least a month.

To collect the magnetic bacteria in the jar, bring the south pole of a magnet up to the outside of the jar opposite and a little above the line where the mud and water meet. Hold the magnet there at least twenty minutes. Then use a pipette to remove a small sample of water from the jar in the area just opposite the magnet. Place a droplet in a depression slide.

To observe the magnetic bacteria with an optical dissecting microscope (the kind usually found in high schools or homes), adjust the microscope light so that the background is dark. Move the magnet from side to side to make the bacteria change direction.

the attracting end of a magnet up to the side of his jar of mud, bacteria would rush toward the magnet. Then he would swoop in with a micropipette and transfer them to another jar. Thanks to their rapid rate of reproduction, he soon had jars full of millions and millions of magnetic bacteria.

Later, Nancy Blakemore used her green thumb and the aseptic techniques she had acquired as a registered nurse to maintain enormous populations of magnetic bacteria for many years. Eventually, she could bring a magnet to the side of a water drop and attract so many magnetic bacteria that they formed a little white crescent, like a thin white fingernail, visible to the naked eye. As Blakemore described it, "As you watch, you'll see them gradually disappear and just sort of dissipate. You'll suddenly blink, and it's gone, and you're not quite sure where it went, but it's gone."

"The second real exciting time for me," Blakemore continued, "occurred when I finally had a chance to prepare my samples for electron microscopy." He had discovered the bacteria with an optical microscope, which reflects beams of visible light off a sample. An electron microscope, on the other hand, would focus electrons on the bacteria and translate the pattern of transmitted electrons into visual information. Because the wavelengths of electrons are about a hundred times smaller than those of visible light, electron microscopes provide greater resolution and sharper images than any optical microscope. Looking through the instrument for the first time, Blakemore identified the tiny rod-shaped bacteria as *Aquaspirillum magneto-tacticum* (later renamed *Magnetospirillum magnetotacticum*). He saw more than the rod shapes of the bacteria, though. Inside each bacterium, he could see the image of a tiny chain of opaque, dark particles (see figure 4.1).

"It blew me away. Inside these bacterial cells were little particles that were actually shimmering, dancing in the electron beam, dark particles similar to what you see in the photographs of them. But in the electron microscope—presumably because of charging effects, they actually lose and pick up intensity. So they're kind of like shimmering, fading in and out, one is dark, one of them is light, then it becomes dark, then one adjacent becomes light, so it's really kind of a little dancing scene. It totally blew me away. And I knew then it was a pretty major find." He began to suspect that the dark particles inside the bacterial cells might be magnetic material.

By this time, Blakemore's adviser at the University of Massachusetts, Professor Ercole Canale-Parola, was advising his student, "This is your discovery. You need to associate yourself with it." So Canale-Parola arranged for Blakemore to spend a summer at Woods Hole Oceanographic Institution on Cape Cod, where the magnetic bacteria had originated. As Blakemore said, "I really appreciated that he didn't grab the project and run with it himself."

When Blakemore was invited to announce the discovery of magnetic bacteria at an important meeting of the American Society for Microbiology in New York City, a University of Illinois professor, Ralph Wolfe, gave him some more good advice. "You really need a motion picture of these in order to convince your audience that they really do move and respond, and that it's not some response that you have to sit and wait for an hour to see or come back after a couple of days to see." Professor Wolfe, the same microbiologist who would

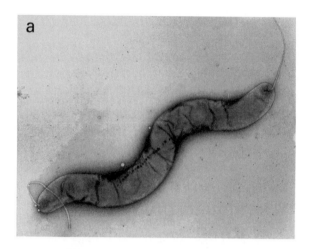

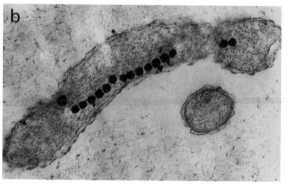

FIGURE 4.1. ELECTRON MICROGRAPHS OF THE MAGNETIC BAC-
TERIUM *MAGNETOSPIRILLUM MAGNETOTACTICUM*.
In view a, *the magnetic particles in are visible as black dots, and two flagella protrude from the ends of the bacterium, which is about 3 micro-
meters long. The magnetic particles in the thin-section, closer view* (b)
*are about 43 nanometers long. Courtesy of N. Blakemore, R. Blakemore,
and R. B. Frankel.*

later advise Carl Woese on his discovery of archaea, loaned Blakemore a 16-millimeter camera.

Wolfe was right. When the film showed Blakemore reversing the
magnet and the bacteria turning and rushing the opposite way, the audience gasped. "Oh, they loved it, they loved it," Blakemore recalled.
"It was incredible. It was really—just *perfect.* Probably everyone in

the audience let out an exclamation. And, in fact, that still happens. If I ever show that film anymore—and I rarely do—you generate an audible response from the audience when they see it."

From then on, as Blakemore puts it, "We were blessed. . . . You'd never get funding if you told someone you wanted to go find magnetic bacteria in mud. But the things we wanted to do we could either do ourselves or do through collaborations. If I approached anyone about the project, they immediately wanted to work on it. You didn't have to do a sales pitch if you wanted to measure this or that. We could always find people who were interested. It was a great interdisciplinary project."

The Magnets

Working with an electron microscope, Blakemore learned that each bacteria contains approximately twenty-two of these little magnets, that the magnets are all roughly the same shape and size, and that they contain iron. Most surprisingly, all the magnets in a bacterium were lined up in a chain that looked for all the world like a tiny compass needle. But what kind of iron mineral was it?

To find out, Professor Canale-Parola urged Blakemore to find a physicist who specialized in magnetism. Richard Frankel, then at the Francis Bitter National Magnet Laboratory at the Massachusetts Institute of Technology and now at California Polytechnic State University at San Luis Obispo, is a self-described "big fan of iron. After more than twenty years, I'm still intrigued with it because the story keeps enlarging. It has an onion-like quality; you add layer upon layer. The story keeps getting deeper and more interesting." After hearing Blakemore talk, Frankel suggested a collaboration, and in 1978, the two started what both remember as "wonderful years."

Their first job was to figure out what makes the bacteria magnetic. Frankel decided to use a technique called Mössbauer spectroscopy, which physicists often use to study molecules containing iron. The process was named for the 1961 Nobel Prize–winning physicist Rudolph Mössbauer.

At one end of his equipment, Frankel placed a source of radioactive cobalt, which was giving off beta and gamma rays as it turned into iron. On the other side of the apparatus, he arranged a gamma-ray detector. In the middle and directly in the path of the gamma rays, Frankel put a small sample of freeze-dried magnetic bacteria.

Whether the gamma rays bounce off the bacteria, pass through them, or are absorbed depends very sensitively on the chemical environment of the iron atoms in the sample. By changing the energy of the gamma rays by only a single part in 100 billion parts, Frankel could ensure that the gamma rays would be absorbed by only one type of iron complex and not by any other. By determining how many gamma rays bounced off, passed through, or were absorbed, and by observing how much he had to change the energy of the rays, Frankel could identify the iron complex in the bacteria.

During the experiment, most of the gamma rays passed through the bacteria and some bounced off. A few, however, were absorbed in such a way that Frankel knew that the compound inside the bacteria had to be the magnetic oxide magnetite. Scientists know magnetite as a complex iron oxide ($FeO + Fe_2O_3$), but sailors knew it for hundreds of years as lodestone. In short, Blakemore's tiny bacteria swim north because they contain tiny magnets directing them toward the North Pole.

Some quick, back-of-an-envelope calculations startled the scientist. Magnetite is crystalline and forms little octahedral or cubic magnets. The bacterium wraps a transparent lipid film around each magnet, making a tiny magnetic packet called a magnetosome. For their size, these magnets contain an enormous amount of iron, more than 1 percent of the bacterium's dry weight. Each magnet contains several million iron oxide molecules. "That's a huge amount," Frankel observed, "a hundred times what a normal organism has." Tiny as they are, magnetic bacteria are the world's most prodigious accumulators of iron.

The bacteria also produce the magnetite at tremendous speeds, considering that a bacterium reproduces every few hours or daily. The bacteria's magnet-making prowess is all the more spectacular because the manufacture of magnetite is normally difficult. In igneous and metamorphic rock, for example, the formation of magnetite requires large amounts of oxygen and extremely high temperatures and pressures.

Armed with his measurements, Frankel spent an evening at home, mulling over the data. Why do all the bacteria's magnets measure between 150 and 200 molecules across? Why does each bacterium have approximately twenty-two magnets? And why are the magnets lined up like a chain? What is the physical significance of that? And how magnetic does the bacterium have to be? "What does it mean?" he kept asking himself.

Sometime that evening, he recalls, "it all hit me at one time like a flash. Remembering some articles I had read, I suddenly realized what it means. It kind of fell into place for me rapidly," Frankel recalled. The frugal little bacterium had discovered the smallest possible magnet that can orient in water. In the language of physics, the bacteria made each of their tiny magnets the size of a single magnetic domain. By now, Frankel was feeling "very high" indeed.

Magnetism occurs only in crystalline solids, which are composed of atoms arranged in a regular and repeated pattern like three-dimensional wallpaper. Electrons are tiny spinning bar magnets, and, in most solids, half of an atom's electrons point in one direction and half point in the opposite direction, canceling one another's magnetism. But some atoms, including the iron atoms in magnetite, have an odd electron that does not have a partner to cancel its magnetism. Special quantum mechanical forces can then compel the odd (unpaired) electrons of many atoms in a small region of space to point in one direction. This small region in which all the atoms have the odd electrons pointing in the same direction is called a "single domain" (see figure 4.2). A single domain is the minimally sized magnet. (More details appear in a sidebar at the end of this chapter and in the glossary.)

A domain is an extremely small magnet. Large and powerful magnets are assembled from many single domains. And in Blakemore's bacteria, the magnets line up in a chain nose to tail, nose to tail.

Finally, Frankel understood. The bacterium's tiny magnets must be strong enough to stay oriented despite the continuous agitation of neighboring atoms. All atoms have thermal energy, that is, random motion from their heat energy, and the thermal motion of surrounding atoms can knock weak magnets out of alignment. A magnet with only a few atoms is quickly overwhelmed, its compass direction flip-flopping rapidly every which way. Weak, wobbly magnets cannot help bacteria navigate.

That evening at home, Frankel suddenly realized that the bacteria's magnets were exactly the right size to resist thermal flip-flopping.

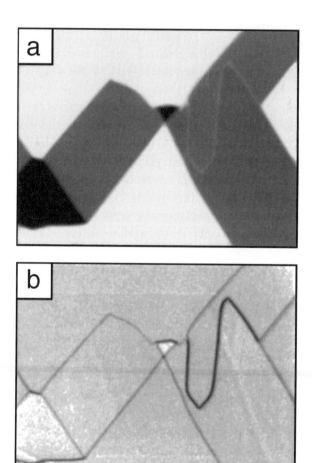

FIGURE 4.2. MAGNETIC DOMAINS.

(a) *The surface of a piece of iron is covered with magnetic domains. Each black, white, and gray region in this polarized electron microscope picture has its spins aligned at a different angle.* (b) *The domain walls between the magnetic domains on the piece of iron shown in a. Courtesy of John Unguris, Robert Celotta, and Daniel Pierce of the Electron Physics Group, the National Institute of Standards and Technology, Gaithersburg, Maryland.*

Magnetic bacteria may be primitive, but they have evolutionarily developed the perfect number, size, and organization of magnetic particles to keep themselves oriented in Earth's magnetic field.

"Given the task of designing an efficiently geomagnetically directed cell, we'd be hard pressed to come up with a superior strategy!" Blakemore observed with awe. First, the bacteria had learned to manufacture magnetite internally and at normal temperatures and pressures. Next, to make the magnets fit inside, the bacteria had discovered the size of a single magnetic domain. Then they lined up the single-domain magnets in a chain of exactly the right length.

The result of all this is a swimming compass needle evolutionarily constructed to direct it along Earth's magnetic field. "And that's the other point about this," Frankel explained. "People read a compass, and then we change direction. But cells don't read; the whole cell is a swimming compass needle. Its orientation is passive. Even if it's not swimming, you'd still see the compass needle orient. When it's dead, it's still oriented. It's always being turned along the magnetic field lines of the Earth."

Magnetic bacteria that died eons ago could have settled to the sea bottom still oriented according to the local magnetic field as it existed at the time. If so, microscopic fossils of magnetic bacteria that drifted to the bottom of the South Atlantic during the Precambrian era about 2 billion years ago could mark the history of Earth's reversing magnetic field. Geologists are studying magnetic bacteria as a possible source of magnetic sediments.

After Blakemore's Cape Cod bacteria orient themselves in Earth's magnetic field, they head north at the stately pace of about 0.1 inch per hour. But what if they are grown under different magnetic conditions? Changing the magnetic fields around bacteria in the laboratory made the little creatures switch directions accordingly.

Bacterial Compasses Everywhere

But what do magnetic bacteria do in nature? Blakemore and Frankel decided they had to visit the Southern Hemisphere to find out whether magnetic bacteria live there too. Compass needles point north and south, but they also point down into Earth's surface as they move toward the poles. The closer the needles get to the poles, the more they dip down along the magnetic field of Earth. This dip, called inclination, increases near the poles and indicates latitude (see figure 4.3). Would Southern Hemisphere bacteria swim north and down like the Cape Cod variety or south and down toward the South Pole?

A mirror image of Woods Hole was needed: a place where everything would be the same as in New England except the direction of the magnetic field. As they searched maps of the planet's magnetic field strengths and magnetic inclination, that is, the angle of a compass needle's dip, the answer became obvious. Australia and New Zealand have temperate climates with roughly the same magnetic field strengths and magnetic inclination as those in New England.

Dick and Nancy Blakemore, Richard Frankel, and Adrianus Kalmijn, a biologist at the Woods Hole Oceanographic Institution, boarded a plane for New Zealand in late December 1979. The team had two weeks to canvass a continent for magnetic bacteria. Blakemore's National Geographic Society grant was big enough only for two-week excursion tickets. Kalmijn and Frankel had each packed a microscope, electric coils, power supplies, and lights.

Once in Christchurch, New Zealand, the team spent a day collecting samples from every pond, river, and stream they could find. Back in their motel that night, they lined up their sample bottles from twenty sites. "By God, we hit them that first night, almost immediately. Of 20 bottles, 12 to 15 had magnetic bacteria," Frankel recalled. Using their microscopes, they confirmed that New Zealand's bacteria swim south and down and that, when the magnet is reversed, they turn around (see figure 4.4).

"We really convinced ourselves. But one other thing we really wanted to do. We wondered, 'Is it the same phenomenon? Do they have magnetite particles inside?'" The next day, they took some of their samples to an electron microscopist at the University of Canterbury in Christchurch. "It wasn't a trivial matter to put a few on the microscope's grid. It's very difficult catching a micron-sized thing with a little pipette, but we hit it. We got a picture of a beautiful chain of particles."

The expedition proved that magnetic bacteria live in the Southern Hemisphere and swim toward the South Pole, but it could not predict what magnetic bacteria would do near the magnetic equator. There the

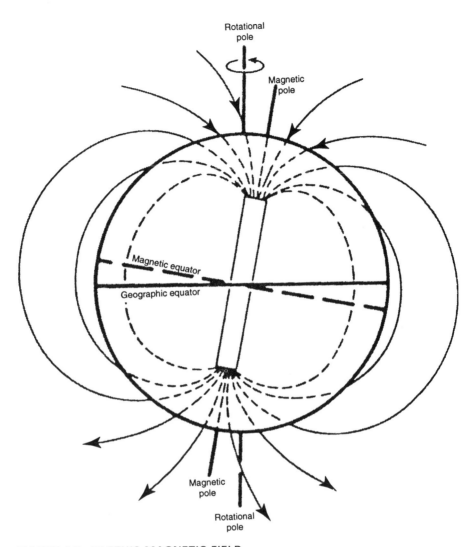

FIGURE 4.3. EARTH'S MAGNETIC FIELD.

The shapes of Earth's magnetic field and the magnetic field surrounding a bar magnet are the same. From motions of the charged particles surrounding Earth, physicists know that lines of force emerge from the magnetic pole in the Antarctic and enter the magnetic pole in the Arctic. The field lines show the direction that a compass needle points. At the magnetic equator, a compass needle lies flat. In the geographic north, a compass needle points (inclines) into Earth. In the geographic south, a compass needle points (inclines) out of Earth. From R. Wiltschko and W. Wiltschko, Magnetic Orientation in Animals *(Frankfurt: Springer-Verlag, 1995). Reprinted with permission.*

Northern Hemisphere   *Southern Hemisphere*

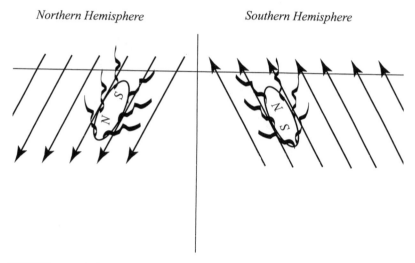

FIGURE 4.4. NORTHERN AND SOUTHERN HEMISPHERE MAGNETIC BACTERIA COMPARED.

No matter what hemisphere they are in, magnetic bacteria swim down into the mud along Earth's magnetic field lines. The compasses grown by bacteria in the Northern Hemisphere are the reverse of those grown in southern hemispheric bacteria.

field is parallel to Earth's surface and does not dip down. The magnetic equator, which is close to the geographic equator, runs through Brazil, Africa, India, and Indonesia. So during the early 1980s, Frankel made two trips to the northeast coast of Brazil.

"Once again, the idea was to keep the environment constant: everything the same except the inclination. And along this coast of Brazil, every time a river gets to the shore, there's a mangrove swamp." So Frankel and a Brazilian colleague, Flavio Torres de Araujo, traveled south, driving out each river road as close to the ocean as possible. Then they hiked out to the mangrove swamp to collect samples.

Northeast Brazil is poor, and many a dirt road ended at a fishing village that, during the 1980s, possessed only one light socket. The socket was generally located in the town's bar, so the scientists would set up their microscope there and the townspeople would crowd around to watch.

"We didn't find many bacteria at the equator, but we found some,"

Frankel recalled. At the equator, without a clear direction from a magnetic field, half of them swam north and half swam south.

"Those were very high times for us," Frankel admitted.

Blakemore's magnetic bacteria had still another surprise in store for their admirers. For years after Blakemore's discovery, scientists assumed that the creatures were stick-in-the-muds, that is, that they made magnets in order to swim down into oxygen-poor bottom mud. But during the mid-1990s, scientists learned how wrong they had been.

It is true that many magnetic bacteria do live in oxygen-poor sediments of fresh water lakes. But they also live in oxygen-poor *water* just below the wave-riffled surfaces of brackish ponds, salt marshes, bays, and oceans. Water there is stratified in horizontal layers that do not mix. In a coastal pond at Woods Hole, for example, the top 6–7 feet (2 meters) are rich with oxygen. But below that, there is almost none, and sulfide diffuses up from the bottom. At least eight species of magnetic bacteria live in this oxygen-poor, sulfide-rich water.

Swimming backward and forward along Earth's inclined magnetic field, bacteria use their propeller-like flagella to move up and down with chemical changes in the water layers. With their flagella rotating clockwise, the bacteria move up into the pond's oxygen-rich zone. If they find too much oxygen there, a chemical switch makes the flagella rotate in the opposite direction to move the bacteria back down to the safe, low-oxygen layers below.

The Missing Link

Although Blakemore's species remains the best-studied magnetic bacteria, others have been discovered. They range in shape from ovals to curved rods or spheres. Each species produces magnetic crystals with a signature size and shape, making it apparent that the bacteria can exert enormous control over their mineralization processes.

Scientists have even found magnetic bacteria that make magnets of iron sulfide instead of magnetite. Discovered in 1990, these bacteria live in sulfide-rich marshes that smell vaguely of rotten eggs. Their individual iron-sulfur magnets are 70 percent less effective than magnetite magnets, but when grouped together in chains, they function almost as well.

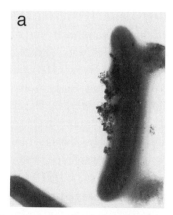

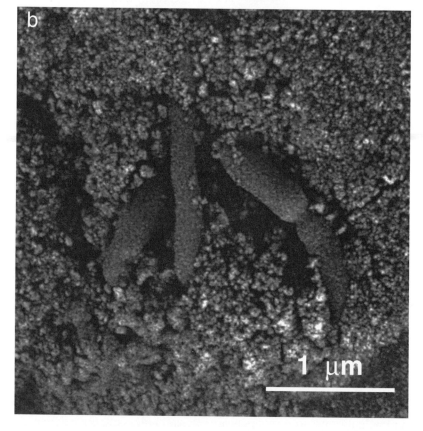

Another type of bacterium was discovered in brackish marshes manufacturing both iron-sulfur magnets *and* the nonmagnetic iron-sulfur complex called fool's gold or pyrite. Fool's gold cannot possibly help the bacteria navigate, but they nevertheless devote enormous amounts of energy to making the two minerals and assembling them in the same chain. Because sulfur can be toxic to a cell, the bacteria may manufacture the compounds necessary to store some of the sulfur out of harm's way.

While collecting microorganisms in Potomac River mud in 1987, Derek Lovley and colleagues then at the U.S. Geological Survey in Washington, D.C., discovered a bacteria, *Geobacter metallireducens,* that takes up ferric iron and then produces *and ejects* large amounts of magnetite without ever becoming magnetic itself (see figure 4.5). By weight, *Geobacter metallireducens* can produce five thousand times more magnetite than an equivalent biomass of Blakemore's *Magnetospirillum magnetotacticum.* Because the former can produce and eject metallic compounds, these bacteria may eventually be used to help purify water polluted with heavy metals.

For many years, biologists and physicists regarded magnetic bacteria as fascinating curiosities. Eventually, however, other scientists also began to suspect that these odd organisms might have played important economic and evolutionary roles. Geologists, for example,

FIGURE 4.5. BACTERIA THAT PRODUCE MAGNETITE WITHOUT BECOMING MAGNETIC THEMSELVES.

A transmission electron micrograph (a) *of GS-15* Geobacter metallireducens *shows magnetic particles (fine-grained magnetite) on the outer left side of the cell. From D. R. Lovley et al., "Anaerobic Production of Magnetite by a Dissimilatory Iron-Reducing Microorganism,"* Nature *330 (1987): 252, fig. 2b. Reprinted with permission from D. R. Lovley and* Nature. *Three bacteria* (b), *one of them dividing, are surrounded by fine grains of magnetite and siderite fluff, which they made. These bacteria, photographed by a scanning electron micrograph, were cultured from samples taken from about 2,000 meters below the surface of the earth. From Liu et al., "Thermophilic Fe(III)-Reducing Bacteria from the Deep Subsurface,"* Science *277 (1997): 1,106, fig. 2a. Copyright 1997 the American Association for the Advancement of Science. Reprinted with permission of T. J. Phelps and* Science.

wondered whether magnetite compasses made by bacteria may have helped create the layers of magnetic material found in the banded iron formations of Australia, Minnesota, and other parts of the world.

During the 1990s, interdisciplinary teams of scientists studying deep, underground formations for the United States Department of Energy discovered living archaea subsisting on rock in Washington State half a mile (800 meters) underground and bacteria living in rock in Virginia 1.7 miles (2.7 kilometers) underground. Magnetite was associated with both these organisms. Like the archaea living in deep-sea vents, the rock-loving archaea and bacteria lived in hot temperatures under pressures three hundred times surface pressures. Taken back to the laboratory, the creatures formed black precipitates, primarily magnetite, within two weeks.

Genetically compared with other life forms, these heat-loving, magnet-forming microorganisms proved to be a missing link between the magnetic bacteria from cool muds on Earth's surface and the microorganisms from hot deep-sea vents, microorganisms that may have been some of the first life on Earth.

Scientists have been able to decipher the complexities of magnetic bacteria largely because they are relatively primitive and simple microorganisms. But vertebrates, which are much more advanced living creatures than bacteria, also wrestled with the problems posed by the oxygen molecules in Earth's atmosphere. More biochemically complex than bacteria, energy-hungry vertebrates used iron atoms to create complex systems to harness, not avoid, oxygen's powerhouse. In doing so, vertebrates built a set of large and remarkable iron-dependent molecules: hemoglobins and myoglobins.

MAGNETISM

Iron atoms are the premier organizers of magnetic effects. If you could shrink to one-trillionth the size of a human and take a seat inside an iron atom of a magnet, what would you see? Every atom—whether it is iron or some other element—is composed of neutrons, protons, and electrons. Each of these particles is a tiny bar magnet with north and south poles and a magnetic field around it. One end of each particle is a north

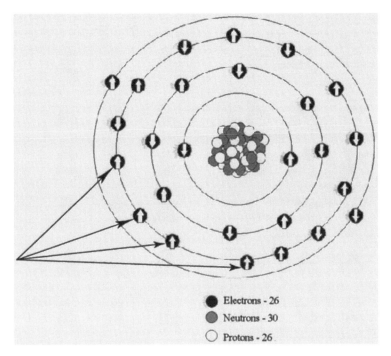

FIGURE 4.6. THE FOUR ELECTRONS (INDICATED BY ARROWS) RESPONSIBLE FOR IRON'S MAGNETISM.

pole, and the other end is a south pole. Opposite poles attract each other, and similar poles repel each other.

Nuclear forces glue the protons and neutrons at the center of the iron atom very tightly together into a ball with a diameter 10^{-14} times shorter than the length of a meter. You would also see the atom's twenty-six spinning electrons occupying a huge area one hundred thousand times larger than the nucleus. Although electrons are negatively charged, they have the same amount of charge as protons.

Perching inside an iron atom, you would see eleven of its electrons spinning clockwise. Eleven more would be spinning counterclockwise. These twenty-two electrons form pairs. Their spins are thus canceled out. The remaining four electrons are unpaired, and they spin in one direction, either clockwise or counterclockwise (see figure 4.6).

Hopping from atom to atom through the magnet, examine each group of four unpaired electrons. Determine which way

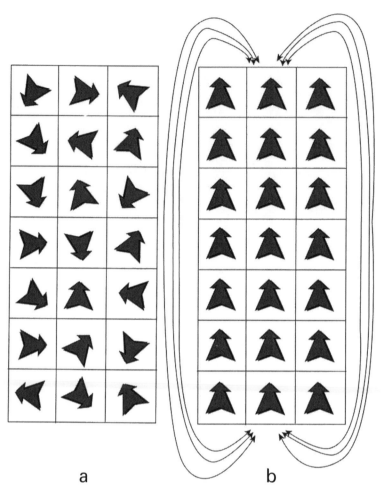

a   b

FIGURE 4.7. MAGNETIC DOMAINS.

Dropping or heating a magnet causes misalignment of the spins of neighboring domains (boxes), and the magnet becomes demagnetized (a). After being placed in the field of a large laboratory magnet and then removed, the magnet is remagnetized (b). The laboratory magnet forced the alignment of all the domain spins and restored the magnetic field.

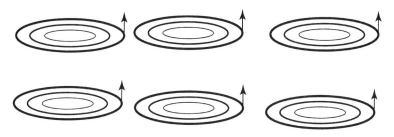

FIGURE 4.8. THE MAGNETIC EXCHANGE INTERACTION.

This figure is simple schematic showing how the ferromagnetic interaction aligns the spins of six neighboring atoms. The arrows indicate the spin of the unpaired electrons in the atoms. The exchange interaction operates so that the unpaired spins on each atom force the other atoms to align their spins. This interaction can exist over a distance of about 1,000 atoms and can form a magnetic domain about 1,000 atoms in diameter. A neighboring domain may have its spin aligned at a slightly different angle.

they are spinning. After visiting approximately one thousand iron atoms, look behind you and you will see that the unpaired electrons are pointing in the same direction. Physicists say that all their spins "agree." They call these regions of agreement "magnetic domains."

Do not stop your tour at a thousand atoms though. Moving beyond the first thousand atoms, you will discover that the unpaired electrons there have slipped out of alignment. But do not give up. Continue your tour and soon you will once again be in another magnetic domain where all the arrows agree. The boundary areas between magnetic domains are called domain walls. Warming a magnet in the sun or an oven disorders the spins and shrinks the regions of agreement. But placing the magnet in a large laboratory magnet realigns the spins and expands the regions of agreement again (see figure 4.7). A special quantum effect called the exchange interaction makes all the unpaired electrons within a thousand-atom area point in one direction (see figure 4.8). This phenomenon, known as ferromagnetism, occurs in only 5 of the 111 known elements: iron, cobalt, nickel, gadolinium, and dysprosium. Physicists do not yet completely understand how the exchange interaction forces unpaired spins of the electrons in neighboring atoms to align. As a result, although scientists can make many ferromagnetic compounds of these elements, they cannot predict what the domain properties of new magnetic compounds will be.

Compressed over geologic time, some iron oxides form lodestone, the fabled magnet of ancient navigators. Depending on iron's crystalline form and the stresses and strains that act upon it, iron's magnetism can vary widely in strength.

5

Hemoglobin and Myoglobin
HARNESSING OXYGEN

Two and a half billion years after oxygen molecules appeared in our atmosphere—long after the iron in the oceans had rusted and most of the planet's iron-loving microorganisms had become extinct—vertebrates evolved by drawing on life's ancient heritage of iron. Thanks to iron's ability to grab and hold on to oxygen, vertebrates could commandeer Earth's new supply of these energy-rich oxygen molecules. This fusion of old and new took place in hemoglobin and myoglobin, the two iron-centered molecules that today stoke human brains and muscles with oxygen.

Hemoglobin, the iron-containing pigment in red blood cells, is an immense and convoluted molecule that evolved about 450 million years ago to carry oxygen from the lungs to the tissues. Hemoglobin and its muscular counterpart, myoglobin, use the same iron-plus-oxygen ingredients that in another form doomed the microbial world to mass extinction and that created vast iron oxide, rust, deposits on planet Earth. But hemoglobin's vast array of eight thousand atoms is elaborately arranged to protect the molecule's four—and only four—atoms of iron from rusting away. These four iron atoms, the workhorses that transport oxygen molecules through the human body, play a far more important role in the body than all the other thousands of atoms in hemoglobin. Without their iron atoms, red blood cells could not function.

Armed with hemoglobin's ability to protect its iron atoms from oxidizing, every vertebrate on Earth could defy the ancient, ironclad rule of inorganic chemistry that iron in the presence of oxygen must rust irreversibly away. Not only does the iron in hemoglobin remain virtually rust free, but iron's hold on oxygen in hemoglobin is also reversible.

The iron in hemoglobin and myoglobin not only grabs and carries oxygen but also releases it in tissues that need more oxygen.

Without their iron-based circulatory systems, humans and other animals would not exist. Without molecular oxygen (O_2) our brains and muscles would not have enough energy for work. Burning food in the presence of oxygen molecules produces eighteen times more energy than burning food without oxygen. Thus iron, which had been a precious metal on anaerobic Earth, became the crown jewel of the planet's new, oxygen-rich atmosphere.

By the time vertebrates evolved a half billion years ago, the second most abundant gas in our atmosphere was oxygen, most of it in the form of molecules (O_2).

To take advantage of O_2's energy potential, most living systems developed a technique for absorbing molecules from the atmosphere. Small aerobic animals, such as insects, could get O_2 directly from the air by diffusion. However, vertebrates, driven by their much greater energy demands, developed a sophisticated method so that a fluid—in their case blood—could transport oxygen molecules throughout the body. Thus, in humans, oxygen breathed in from the atmosphere flows through the trachea and bronchial tubes to the lungs. From there, the hemoglobin in blood delivers the oxygen to the capillaries and thence, via myoglobin, to the muscles and other organs and finally to the mitochondria. There, glucose and sugar are "burned" in the presence of massive quantities of molecular oxygen.

Before vertebrates could use their blood to transport oxygen, they had to solve a critical problem. The gaseous oxygen we absorb from the atmosphere does not dissolve well in blood. Blood can dissolve only enough O_2 to supply 2 percent of a human body's needs at rest and an even smaller percentage of an active body's requirements.

To augment this paltry supply of dissolved molecular oxygen, our red blood cells contain an intricate iron-based system using hemoglobin molecules. In molecular terms, hemoglobin is a veritable shovel. Its iron atoms load up with oxygen molecules in the lungs and carry them through the blood stream to capillaries in the tissues. Each human red blood cell contains approximately 280 million of these shovels, enough to give blood the consistency of glue.

Iron's activities in the blood are conveniently color coded. Oxygen-enriched hemoglobin is scarlet as it races from the lungs through the arteries to capillaries in the tissues and organs. There the hemoglobin releases its oxygen. Turning bluish red, oxygen-depleted blood picks up carbon dioxide (CO_2) and hauls it back to the lungs via the veins.

Structure and Function *my blue legs!*

Hemoglobin is composed of four long chains of amino acids, the building blocks of proteins and peptides. Each chain includes one iron atom that can transport one oxygen molecule (O_2). The system is quite efficient. Red blood cells loaded with fully oxygenated hemoglobin can carry approximately eighty times more oxygen than blood can carry as a dissolved gas.

The body of the average healthy adult contains 3 or 4 grams of iron, about one-tenth of an ounce, roughly the weight of a penny. Between 60 percent and 70 percent of this iron is in hemoglobin. The rest is in ferritin (the iron-storage compound described in chapter 3), in myoglobin, and in other iron compounds.

Iron's cycle within the human body starts when iron is absorbed from food in the small intestine. From there, the iron passes, via the blood plasma, to the bone marrow. In the marrow, the iron is used to complete the hemoglobin molecules, which are, in turn, incorporated into red blood cells. If any step in the process goes awry, disaster can ensue. For example, when lead accumulates in bone, it interferes with the enzyme that inserts the iron into hemoglobin. Or, in another example, if the walls of the red blood cell break and hemoglobins spill out, the ferrous iron quickly reacts with oxygen in the body and turns irreversibly to ferric oxide. Hemoglobin thus exposed to oxygen in the atmosphere forms a brown pigment called ferrihemoglobin or methemoglobin. The pigment makes rusty brown stains. The oxygen in ferric oxide can no longer be used by the body. Several blood diseases lead to the oxidation of iron atoms in hemoglobin inside red blood cells; these diseases quickly destroy the red blood cells and prevent them from transporting oxygen through the body. Anemia can result, and if more than 5 percent of the hemoglobin in blood is transformed into methemoglobin, the skin can look slightly bluish, a condition called cyanosis.

Healthy blood cells live for approximately four months. When red blood cells die and disintegrate, their iron is released and recycled into new hemoglobin to begin the cycle anew. During this same period, the body absorbs and excretes one-tenth of a gram of iron daily, and hemoglobin recycles more than a gram of the metal into new red blood cells.

When the English scientist Max Perutz started deciphering hemoglobin's structure in 1937, he thought that the job looked straightforward. But when he chose the molecular structure of hemoglobin as his doctoral thesis project, using crystallography for his analysis, fellow students at the University of Cambridge smiled with pity at his folly. Crystallography—the x-ray analysis of crystals—was still widely regarded as an unproved black art. The most complex molecule that had been determined by crystallography was composed of only fifty-eight atoms. But Perutz thought crystallography offered the only chance, albeit an extremely remote one, of locating the thousands of atoms in a hemoglobin molecule. The first x-ray diffraction pictures of protein crystals had been taken three years earlier. The importance of nucleic acids had not yet been discovered, and the "secret of life" was thought to lie in the structure of proteins, which provide amino acids for the growth and repair of plant and animal tissues. In the meantime, when Perutz's thesis adviser, J. D. "Sage" Bernal, and Dorothy Crowfoot Hodgkin discovered that protein molecules, despite their large size, have highly ordered structures, the news caused a sensation.

Eventually, the development of computers and new crystallographic techniques enabled Perutz to refine a three-dimensional map of hemoglobin enough to reveal the molecule's overall structure. In the summer of 1959, nearly twenty-two years after Perutz started his thesis, the structure of hemoglobin emerged at last, revealing four coiling amino acid chains. Fortunately, as he liked to joke, his graduate school examiners had granted him his degree long before that.

As Perutz and his colleagues plotted their data onto contour maps, they realized that each of hemoglobin's four chains closely resembles the single chain of myoglobin. John C. Kendrew had discovered myoglobin's structure at Cambridge University two years earlier (see figure 5.1). Kendrew would share the Nobel Prize for Chemistry with Perutz in 1962. In fact, hemoglobin's four chains are almost identical to one another and to myoglobin's single chain.

Every animal hemoglobin molecule—whether from horse, human,

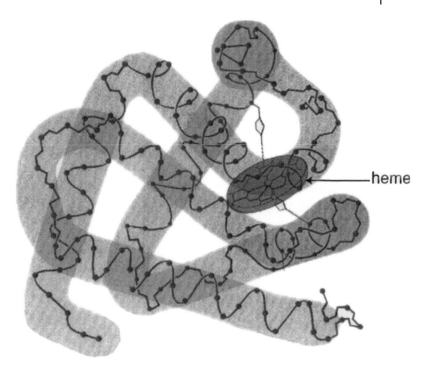

heme

FIGURE 5.1. MYOGLOBIN'S STRUCTURE.
Myoglobin, the protein that colors muscle red, has only one chain, which is similar to a hemoglobin beta chain. The chain winds around the heme, at the center of which is one iron atom where the oxygen molecule binds. From L. Stryer, Molecular Design of Life *(New York: W. H. Freeman, 1989). Copyright © 1989 by Lubert Stryer. Reprinted with permission of W. H. Freeman & Company.*

bird, or crocodile—has basically the same four-chain structure. In round numbers, each of hemoglobin's four chains is composed of two thousand atoms coiled into eight helical sections. About half the atoms are hydrogen, and the rest are carbon, nitrogen, oxygen, and sulfur. There is only one iron atom per chain. Most of the atoms in a chain are built into amino acids, the globin part. Two of the hemo-globin chains, called alphas, contain 141 amino acids, while the other two chains, called betas, have 145 (see figure 5.2).

Each hemoglobin chain wraps like swaddling clothes around an iron-centered group of atoms. Removing the iron group relaxes the chains and makes them unfold; reinstating the iron group makes the chains close back up into a taut, biologically active molecule. Changing the order of even one amino acid at a sensitive point can make the chain fold improperly and malfunction.

Analyzing hemoglobin, Perutz was surprised to see that however the chains folded, they always formed a neat and almost spherical molecule. He was also astonished by the fact that every hemoglobin molecule consistently folds into a stable structure. He concluded that the hemoglobin molecule had evolved so that it could hold the iron-centered group of atoms called "the heme" rigidly in place.

A heme is a disk composed of an iron atom surrounded by a so-called porphyrin ring of carbon, nitrogen, and hydrogen atoms (see

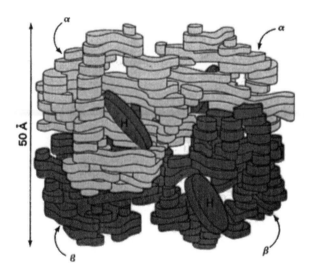

FIGURE 5.2. HEMOGLOBIN'S STRUCTURE.

Hemoglobin, a huge and convoluted molecule, is composed of four amino acid chains, two alphas and two betas, that wrap protectively around disk-like hemes (labeled H in the diagram). Two of the hemes are clearly visible in the front of the diagram, and two are in back. At the center of each disk sits a single iron atom where an oxygen molecule can dock. Hemoglobin is roughly 50 angstroms long. From L. Stryer, Molecular Design of Life *(New York: W. H. Freeman, 1989). Copyright © 1989 by Lubert Stryer. Reprinted with permission of W. H. Freeman & Company.*

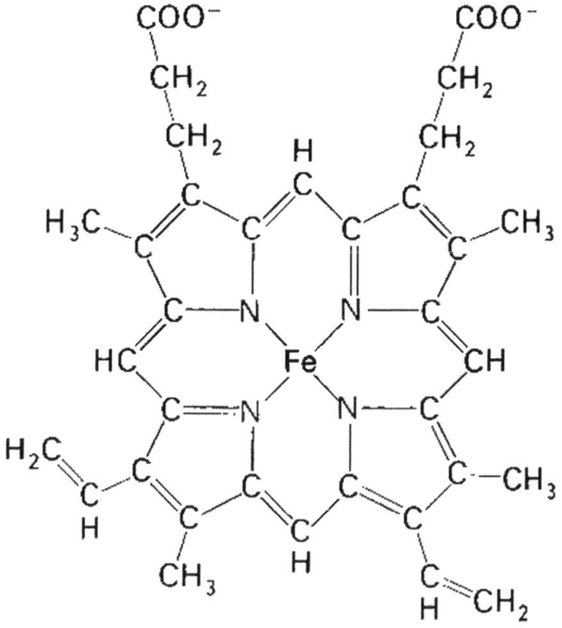

Heme
(Fe-protoporphyrin IX)

FIGURE 5.3. THE HEME.
Oxygen binds to an iron atom at the center of the heme. The disk-like heme is composed of a porphyrin ring of carbon, nitrogen, and hydrogen atoms that surround the iron atom. The lines in the diagram represent the bonds between the atoms.

figure 5.3). Perutz thought the iron atom looked like a jewel set in a ring. Combining the word "heme" with "globin," the term for the protein chain component of the molecule, gives hemoglobin its name.

The heme is one of the essential pigmented molecules of life, together with chlorophyll and vitamin B_{12}. All three are built of a single metal atom surrounded by a porphyrin-like ring of atoms. Chlorophyll has a magnesium atom in the center of its porphyrin, whereas B_{12}, the anti-pernicious anemia factor, has a cobalt atom in the middle. Because of the differing interactions between their proteins and their metal, plants appear green, and vitamin B_{12} appears deep red.

Around the heme, unique chemical conditions created by the folded chains inhibit oxidation reactions at the iron site. Each hemoglobin chain curls protectively around its iron center and embeds the heme in a helical fold or pocket on the surface of the molecule. Only a small portion of the heme protrudes from its pocket. However, if water enters the pocket, the chain unfolds, and free oxygen molecules dissolved in the blood oxidize the iron to rust.

Hemoglobin offered scientists the first direct demonstration that chemical changes involving proteins are also accompanied by structural and functional changes. Proteins perform a wide variety of jobs depending on their shape when tightly folded. Hemoglobin became the best understood of these proteins. As such, hemoglobin is a touchstone for the development of new theories and experimental techniques and a keystone in the history of modern structural molecular biology.

Cooperation

Entering the alveolar capillaries of the lungs, a hemoglobin waits to pick up oxygen. Without oxygen, the molecule is tense, its chains held tightly together by chemical bridges and electrical forces. The hemoglobin's iron atoms sit in their hemes. The hemes are tilted back slightly toward the chain's exterior.

After the oxygen molecules cross the membrane of a red blood cell, they collide over and over again with the waiting hemoglobins. Access to the iron atoms and their empty binding sites is difficult because the sites are protected inside the pockets of the hemoglobin chains. But some oxygen molecules do find their way through the hemoglobin's maze of alpha and beta chains. The higher the oxygen concentration, the more frequent the rate of collision and binding. In the oxygen-rich lungs, hemoglobin quickly loads up with oxygen.

Pure chance determines which iron site will be the first to bind an oxygen molecule. But hemoglobin's chains are unusually sensitive to their environment. Once an oxygen molecule docks onto an iron site, the hemoglobin's rigid structure relaxes. Hemoglobin's opening for the first oxygen molecule makes it seventy-five times easier for the remaining three iron atoms in the chains to bind an oxygen molecule. The progressive interaction between alpha and beta chains is called "cooperativity" because the binding of the first oxygen stimulates the other three chains to relax. Cooperativity boosts the probability that the other iron sites will bind oxygen. Without this cooperative effect, the last three oxygens would experience roughly as much trouble docking onto the iron as the first one did. Cooperativity gives people and other vertebrates exquisite control over their oxygen supplies.

How did this cooperativity evolve? Beginning about 450 million years ago, the hemoglobin gene diverged from myoglobin, and hemoglobin acquired alpha and beta chains. By clustering pairs of chains together so their heme sites could interact, hemoglobin could react with increasing speed to changes in its chemical environment. What limited the number of hemoglobin's chains to four? Was it the stability of the molecule? Would it fall apart if it got too big? Perhaps if hemoglobin acquired too many chains, switching back and forth between oxygen transport and oxygen dumping might become too slow and cumbersome. Four chains may have been the most efficient compromise between cooperativity and reversibility. Theoretical chemists and physicists use quantum mechanics to study the stability of these chains.

Because of this cooperativity between chains, the heme and oxygen work together efficiently. Loosening up the chains increases the probability that each hemoglobin will travel to the capillaries fully loaded with oxygen. Hemoglobin molecules leave the lungs 97 percent saturated with molecular oxygen.

When an oxygen molecule approaches an iron site, conditions around the iron enable it to grab the oxygen in a uniquely physical way. As a result, the two do not combine to form an iron oxide, the familiar rust. How is the feat accomplished? Before the oxygen docks, the iron is in its ferrous valence state, that is, the neutral atom has effectively transferred two of its outer electrons to neighboring amino acids. Normally, when exposed to oxygen, the ferrous iron would quickly lose another outer electron and become oxidized to its ferric state as rust.

Surprisingly, when O_2 docks onto the iron in hemoglobin, a third electron does *not* move from the iron to the oxygen. Instead, the electron stays with the iron, and two electrons at the iron site content themselves with reversing the direction of their spins. It is even more amazing that this electron spin reversal is the most important chemical change that occurs when the oxygen docks onto the iron. The iron remains in the same chemical state, the ferrous (Fe^{++}) state, and is not oxidized to the ferric (Fe^{+++}) state by the presence of oxygen.

Before the oxygen pops onto the iron, the iron nestles comfortably below the flat porphyrin ring. When the oxygen docks and the electrons flip their spin, the iron is pulled ever so slightly into the center of the heme (see figure 5.4).

Dumping

The iron atoms in hemoglobin are far more than just oxygen carriers, however. They are also oxygen "dumpers" or, in technical terms, reversible oxygen transporters. Because of this, oxygen can slip off the iron into tissues as easily as it slips onto the iron in the lungs. This easy-come-easy-go chemical reaction is called reversible oxygenation.

FIGURE 5.4. DOCKING.
An oxygen molecule, the dark dumb-bell in the diagram, docks onto the iron atom (Fe) at the center of the heme, which is viewed from the side here. Reprinted with permission of the Protein National Data Bank.

Thus, the cooperativity between hemoglobin's four chains is probably the single most important feature that hemoglobin evolved over its 450-million-year development. By giving vertebrates the ability to load and unload oxygen fast, cooperativity helped give them the power to spring into action, to pounce on prey, or to evade enemies fast.

Myoglobin, the protein that colors muscle red, is hemoglobin's smaller and more primitive cousin. It binds oxygen even more tightly than hemoglobin and dumps a third of it all at once into oxygen-hungry exercising muscles. But because myoglobin consists of only one chain, it cannot moderate its response to the muscles. If hemoglobin had evolved as a one-chain oxygen-delivery system, it would dump into the muscles only a small fraction of the oxygen it carried.

Hemoglobin's remarkable sensitivity to changes in oxygen pressure is part of the secret behind a vertebrate's ability to give sustained chase or to perform hard physical labor for long periods of time. Figure 5.5 shows how much oxygen can be loaded on and dumped off hemoglobin under different oxygen conditions. The plot shows how hemoglobin can dump large amounts of oxygen into the body over a very narrow range of oxygen pressures, that is, when working tissues are oxygen hungry.

HEMOGLOBIN AND MYOGLOBIN DELIVER OXYGEN

Figure 5.5 depicts a broad range of oxygen pressures in the body. In a healthy person at sea level, the pressure of the oxygen molecules in the lungs measures 110 mm Hg (point A on figure 5.5). Outside the body, of course, the total atmospheric pressure at sea level is 760 mm Hg, of which 150 mm Hg is due to oxygen, because nitrogen, hydrogen, and other gases are also present. At point A, the hemoglobin molecules leaving the lungs in the red blood cells are 97 percent saturated with oxygen. This corresponds to 97 percent oxygen saturation on the y axis. At point B, the oxygen pressure in the tissues of a mildly exercising person is about 40 mm Hg, and the hemoglobin unloads about 25 percent of its oxygen. Note that on the y axis this corresponds to 75 percent oxygen saturation. At this point, the

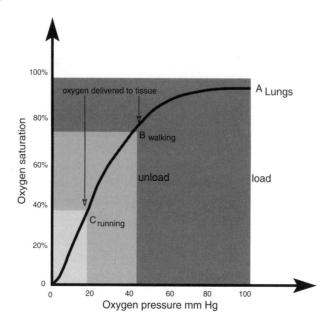

FIGURE 5.5. HEMOGLOBIN DELIVERS OXYGEN.
The x *axis, labeled "Oxygen Pressure," depicts oxygen pressures in both the lungs and the tissues. The* y *axis, labeled "Oxygen Saturation," depicts the percentage of the hemoglobin molecules in the red blood cells that have oxygen bound to them. From G. Benedek and F. M. H. Villars,* Physics, *with Illustrative Examples from Medicine and Biology,* vol. 1, Mechanics, *2d ed. (New York: Springer-Verlag, 2000). Reprinted with permission.*

hemoglobin leaves the capillaries with 75 percent of its sites still loaded with oxygen.

As a person increases the rate of exercise, the muscles call for more oxygen and the oxygen pressure in the tissue falls to 20 mm Hg (point C). Hemoglobin responds to this increased need by unloading 35 percent more of its oxygen. This corresponds to almost 40 percent oxygen saturation on the *y* axis. Thus, when this hemoglobin starts its return journey to the lungs, it has dropped a total of 60 percent of its oxygen.

Figure 5.6 contrasts the oxygen affinity of hemoglobin and myoglobin. Note that at all pressures the oxygen affinity of myoglobin is much greater than that of hemoglobin. The diagram shows that at all pressures, oxygen moves from hemoglobin in the capillaries to myoglobin in the muscles. The arrow in the figure illustrates oxygen's delivery from hemoglobin to myoglobin. Note that the shape of the curves reflects this difference in their binding power. Hemoglobin's curve has an elongated S shape, whereas myoglobin's rises steeply and then flattens out.

In normal non-exercising tissues, only about 10 percent of the myoglobins would dump their oxygen (see figure 5.7). Even when oxygen pressures sink between 20 mm Hg and 15 mm Hg, only about 10 percent more of the myoglobin would dump oxygen. But when oxygen pressure drops below 10 mm Hg, myoglobin comes into its own. There fully 30 percent of the myoglobins release their oxygen to the depleted tissue.

Hemoglobin's final job is to sweep up the carbon dioxide (CO_2) produced by the mitochondria's biochemical furnaces and

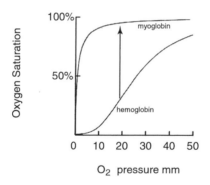

FIGURE 5.6. HEMOGLOBIN COMPARED TO MYOGLOBIN.

This figure contrasts the oxygen affinity of hemoglobin and myoglobin. Note that at all pressures the oxygen affinity of myoglobin is much greater than that of hemoglobin. The diagram shows that at all pressures, oxygen moves from hemoglobin in the capillaries to myoglobin in the muscles. The arrow in in the figure illustrates oxygen's delivery from hemoglobin to myoglobin. Note that the shape of the curves reflects the difference in their binding power. Hemoglobin's curve has an elongated S shape, whereas myoglobin's rises steeply and then flattens out. From L. Stryer, Molecular Design of Life *(New York: W. H. Freeman and Co., 1989). Copyright © 1989 by Lubert Stryer. Reprinted with permission of W. H. Freeman & Company.*

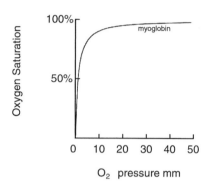

FIGURE 5.7. MYOGLOBIN DELIVERS OXYGEN.

In normal non-exercising tissues, only about 10 percent of the myoglobins would dump their oxygen. Even when oxygen pressures sink between 20 mm and 15 mm, only about 10 percent more of the myoglobin would dump oxygen. But when oxygen pressure drops below 10 mm, myoglobin comes into its own. There fully 30 percent of the myoglobins release their oxygen to the depleted tissue. From L. Stryer, Molecular Design of Life *(New York: W. H. Freeman and Co., 1989). Copyright © 1989 by Lubert Stryer. Reprinted with permission of W. H. Freeman & Company.*

carry it back to the lungs. The hemoglobin in red blood cells carries between 20 and 30 percent of the carbon dioxide transported through the veins. The blood plasma takes the rest. But iron plays no role here; CO_2 does not dock onto iron sites.

To visualize the dramatic evolutionary advantage that hemoglobin's sensitive system of reverse oxygenation gave to vertebrates, imagine a prehistoric hunter out scouting for food. When hemoglobin reaches the oxygen-needy capillaries of the hunter's muscles and organs, it releases its oxygen in stages, holding back some oxygen in reserve for future contingencies. Before the hunter starts working, for example, the oxygen pressure in the capillaries is about 40 mm Hg, and hemoglobin unloads the oxygen from 25 of every 100 iron sites (point B in figure 5.5). But when a woolly mammoth appears on the horizon and the hunter runs toward it, the oxygen pressure in his capillaries drops to 20 mm Hg. At this point, his hemoglobin unloads the oxygen from another 35 sites (point C in figure 5.5).

As the hunter chases the mammoth and the oxygen pressure in his muscles plummets, hemoglobin's ancestral cousin comes to the rescue. Myoglobin, also called "muscle blood," is sometimes thought of as a storage protein. But it is specially equipped to do more than just hold oxygen. It can also release large amounts of oxygen to the running hunter's muscle cells when oxygen pressures in the cells drops extremely low.

Although hemoglobin's four chains work together to react swiftly to subtle chemical changes around them, myoglobin is stable. It binds oxygen even more tightly than hemoglobin. In a normal non-exercising or moderately exercising hunter, only about 17 percent of the myoglobin would dump its oxygen. But when oxygen pressure drops below 10 mm Hg during strenuous exercise, myoglobin comes into its own. There roughly a third of the myoglobin molecules release their oxygen to the depleted capillaries, allowing our panting hunter to hurl his spear and fell the mighty mammoth.

High Altitudes

Like heavy exercise, high altitude also challenges hemoglobin's ability to deliver enough oxygen when needed. More than 10 million people live 12,000 or more feet above sea level, and lowlanders travel there to ski and climb mountains. Yet a breath of air taken at high altitude contains less oxygen than one taken at sea level (see tables 5.1 and 5.2). At the summit of Mount Everest—the highest place on Earth at 29,028 feet above sea level—the blood leaves the lungs with

TABLE 5.1. Oxygen Saturation of Arterial Blood at Various Altitudes

Altitude	Oxygen saturation (%)
Sea level	97
10,000 feet (3,048 meters)	90
20,000 feet (6,096 meters)	70
30,000 feet (9,144 meters)	20

Note: Because atmospheric pressure declines at higher altitudes, the oxygen saturation of arterial blood also decreases. This table shows the drop in oxygen saturation as arterial blood leaves the lungs at different altitudes.

TABLE 5.2. Predicted Decline in Aerobic Power
at Increasing Altitudes

Location (altitude)	Decline in aerobic power (%)[a]
New York City (sea level)	threshold
Denver, Colorado (5,280 feet; 1,609 meters)	5
Mexico City (7,320 feet; 2,240 meters)	8
Quito, Ecuador (9,350 feet; 2,850 meters)	16
Pikes Peak, Colorado (14,110 feet; 4,301 meters)	29
Denali (Mount McKinley), Alaska (20,320 feet; 6,194 meters)	48
Mount Everest (29,028 feet; 8,848 meters)	supplementary oxygen required

Note: As people climb to higher altitudes, they cannot exercise as efficiently because of lower oxygen concentrations in arterial blood when it leaves the lungs. This table shows typical declines in aerobic power at several major cities and mountains.
[a]Aerobic power is the maximum amount of oxygen that can be consumed during one minute of exercise.

only 20 percent of its iron sites binding oxygen. Few climbers have reached the peak without supplementary oxygen.

Not only do people at high altitudes inhale less oxygen, but they also have trouble transferring it to their muscles. This double whammy presents difficulties for even expert climbers. Normally, the big difference in oxygen pressure between the blood plasma and the capillaries of exercising muscles drives oxygen into the myoglobin. Above about 10,000 feet, however, the difference in these cellular oxygen pressures is much smaller, so the rate of diffusion into the muscles slows down.

Travelers arriving at high altitudes adapt to the low oxygen pressure in three consecutive steps. First, they immediately increase their respiration by breathing faster and more deeply. Unfortunately, this hyperventilation makes them exhale more CO_2, which dissolves in body fluids to form an acid and upsets the blood's normal acid-base balance. The blood becomes increasingly alkaline. Acetazolamide, a prescription drug marketed as Diamox, helps the kidneys reacidify the blood.

The second step in acclimation occurs within hours of a person's arrival at high altitudes, when a special phosphate molecule swings into action and helps out. Glycolysis, the burning of sugars in the mitochondria, fuels hard exercise and steps up the production of the molecule 2,3-DPG. Technically known as 2,3-diphosphoglycerate, it decreases hemoglobin's affinity for oxygen. The 2,3-DPG binds directly to the hemoglobin's central cavity close to the iron atoms. It binds so strongly that it displaces oxygen from the iron sites. Strenuous exercise produces large amounts of 2,3-DPG and drives oxygen from the hemoglobin onto the myoglobin molecules for use in the muscles.

Logically, the body could get more oxygen at higher altitudes by increasing either the number of hemoglobins in the red blood cells or the number of blood cells. In practice, the first alternative is impossible because a normal red blood cell cannot hold any more hemoglobin molecules than it already does.

To a limited extent, however, the body can take a third step toward acclimation to high altitudes by producing more red blood cells. The process has a number of drawbacks. It takes three or four weeks. Even people who become fully acclimated to high altitudes can produce only 20 percent more red blood cells than normal. Furthermore, adding more red blood cells to the blood makes it more viscous and therefore harder for the heart to pump through the body. Thus, it is unclear how much increasing the number of blood cells helps acclimation to high altitude. Studies of indigenous populations in the Andes show that the oxygen affinity of their hemoglobin is not significantly better than that of people who live at sea level. Evidently, 2,3-DPG helps visitors adapt to high altitudes better than increasing the blood supply does.

Athletes at low altitudes have also tried to increase oxygen transport to their muscle. During the 1980s, a number of long-distance bicyclists, cross-country skiers, triathletes, speed skaters, and runners experimented with blood doping, also called blood packing or induced erythrocythemia. First, some of the blood was removed and stored. Second, after the body naturally replenished its supply of red blood cells, the stored blood was transfused back into the athlete. In a high-tech version of blood doping, some bicyclists have taken a hormone, erythropoietin (EPO), that stimulates the bone marrow to produce more red blood cells. EPO, which the kidneys and liver produce naturally, was bioengineered for patients with anemia caused by kidney disease, AIDS, or cancer. Because EPO makes the blood thicker, athletes who use it risk having strokes and heart attacks. EPO has been linked anecdotally to two dozen deaths in cycling alone. The drug is hard to detect because it is so similar to its naturally occurring counterpart. Both blood doping and EPO have been banned by the International Olympic Committee because they carry health risks and give those who use them an unfair competitive advantage.

A Gallery of Specialized Hemoglobins

The standard garden-variety vertebrate hemoglobin is a molecular marvel, but it is not the only way to oxygenate a living organism. Over millions of years, creatures have evolved other variations to meet special oxygen-transport needs. Here are a few specialized and variant hemoglobins that help their hosts survive and lead active lives in unusual environments.

Embryonic and Fetal Hemoglobin

How does a developing human embryo or fetus survive the watery environment of the uterus? How does an embryo or fetus absorb oxygen before its lungs function? The answer to these two questions is that the mother's blood and the fetus's blood flow through the placenta in separate capillaries, allowing oxygen to diffuse from maternal to fetal capillaries.

In all, three kinds of hemoglobin form at different times to oxygenate the fetus as it develops inside the uterus. The key to their success is that the first two kinds bind oxygen much more powerfully than adult hemoglobin can. Thus, oxygen moves from the mother to the fetus (see figure 5.8).

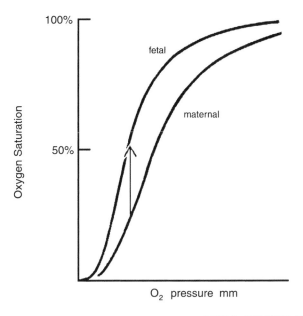

FIGURE 5.8. MATERNAL HEMOGLOBIN DELIVERS OXYGEN TO FETAL HEMOGLOBIN.

The curve for fetal hemoglobin lies above the curve for maternal hemoglobin at all oxygen pressures, indicating that fetal hemoglobin binds oxygen more tightly than does maternal hemoglobin. Therefore oxygen travels from the mother to the fetus, as indicated by the arrow in the diagram. From L. Stryer, Molecular Design of Life *(New York: W. H. Freeman, 1989). Copyright © 1989 by Lubert Stryer. Reprinted with permission of W. H. Freeman & Company.*

During the first eight weeks after conception, the embryo's red blood cells contain a special type of hemoglobin, composed of two so-called zeta chains and two epsilon chains. The sequence of the amino acids in these chains differs from that of adult hemoglobin. After eight weeks, a genetic switch stops the production of this early form of hemoglobin, and the fetus begins making a transitional form of hemoglobin called fetal or gamma hemoglobin. Fetal hemoglobin consists of one pair of gamma chains (composed of their own unique amino acid sequence) and one pair of adult alpha chains. This transitional gamma-alpha hemoglobin accounts for most of the hemoglobin in the fetus until just before birth, when the production of beta

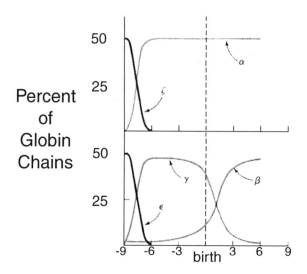

FIGURE 5.9. THE DEVELOPMENT OF ADULT HEMOGLOBIN FROM CON-
CEPTION TO INFANCY.

*During the first eight weeks after conception, the growing embryo stops
making zeta chains (the dark lines in the upper half of the diagram) and epsilon chains (the dark lines at the lower half). At the same time, the embryo
steps up its production of alpha and gamma chains. Close to delivery, the
production of gamma chains drops off and the production of beta chains increases. Six months after birth, the infant has adult hemoglobin: 50 percent
alpha chains and 50 percent beta chains. From L. Stryer,* Molecular Design
of Life *(New York: W. H. Freeman and Co., 1989). Copyright © 1989 by
Lubert Stryer. Reprinted with permission of W. H. Freeman & Company.*

chains accelerates rapidly. An infant does not have fully adult hemoglobin—with equal numbers of alpha and beta chain pairs and a few
remaining gamma chains—until twenty-six weeks after birth (see
figure 5.9).

This late production of beta chains has important implications for
individuals who carry genes that cause severe forms of thalassemia,
a group of blood diseases that underproduce alpha or beta chains. Abnormal hemoglobins may not be detected in early blood samples
taken from the fetus, and genetic tests may be necessary.

Even as adults, some individuals with aplastic anemia, leukemia,
sickle cell disease, and some types of thalassemia can have abnormally

high levels of fetal hemoglobin. Because of fetal hemoglobin's affinity for oxygen, some people with sickle cell disease or thalassemia can escape some of the damage usually caused by those conditions.

Horseshoe Crabs and Marine Worms

Hemoglobin and myoglobin are not the only molecules that can pick up and drop oxygen wherever needed. At the time oxygen was entering Earth's atmosphere, organisms tried a variety of coping techniques. Of these ancient mechanisms, two remain.

A protein molecule called hemerythrin transports oxygen in rubbery marine worms called "peanut worms" and in the clam-like creatures called "lampshells." At the center of this hemerythrin oxygen transporter sit two iron atoms, instead of hemoglobin's four.

Many mollusks and arthropods—including horseshoe crabs—depend on a molecule called hemocyanin. Here the center of the oxygen-transporting protein has two copper atoms instead of one iron atom. Horseshoe crabs, which are arthropods related to the scorpions, spiders, and extinct trilobites, are often called living fossils. They swam on their backs and plowed through mud when dinosaurs ruled Earth 190 million years ago. In fact, horseshoe crabs are even older than dinosaurs. They evolved at least 540 million to 600 million years ago and were a dominant group of marine animals for about 400 million years. Four species remain, so their copper-centered reversible oxygen transporters have proved their worth over time. Nevertheless, these oxygen transporters are larger, enormously more complex, and less efficient than iron-centered hemoglobins. A horseshoe crab must use approximately ten thousand atoms to grasp and hold a single molecule of oxygen, whereas hemoglobin can carry four oxygen molecules with only eight thousand atoms.

Crocodiles

Despite their fearsome jaws and teeth, crocodiles often rely on their hemoglobin to help kill prey. Crocodiles can submerge up to two hours without taking a breath. So they often kill of their victims by holding them underwater until they drown.

Crocodiles can perform this breathless feat because twelve of the amino acids in their hemoglobin differ from those in human hemo-globin. Thus, if we could change a few of our amino acids we too could remain underwater for extended periods of time. The changes in amino acids—slight as they are—dramatically alter the croco-dile's ability to dump oxygen into the tissues. When the animal holds its breath and carbon dioxide accumulates in its blood, this excess carbon dioxide ultimately forces the crocodile hemoglobin to release a large fraction of its oxygen into the tissues. Japanese scientists who were making and studying combinations of human and crocodile hemoglobins in Cambridge, England, in 1995 isolated the functional set of amino acids in the crocodile hemoglobin that enables the ani-mal to submerge so long. Not surprisingly, they named the protein "Scuba."

Deep-Diving Mammals and Birds

Like yo-yo's, female northern elephant seals dive for months nonstop, day and night, taking only two or three minutes between dives to catch their breath at the surface. When they dive, they can stay under for an hour and go deeper than any mammal known: up to four-fifths of a mile down.

Foraging for hake and squid off the continental shelf, they avoid the ocean surface where white sharks hunt. Female northern elephant seals come on land only to copulate, give birth, lactate, and wean their pup. Biologists think that males, which are five to ten times larger than females, may dive even deeper. How do northern elephant seals get oxygen? They rely on vast quantities of oxygen stored not in their lungs but in their blood and muscles. An elephant seal has two and a half times more blood than a human of comparable size. Its blood is viscous with extra blood cells. Its muscles are enormous and packed with huge supplies of myoglobin. Myoglobin, which binds oxygen more tightly than hemoglobin, is three to ten times more concen-trated in northern elephant seals than in humans.

Other marine animals also use oxygen stored in their blood and muscles. A Weddell seal, for example, stores 95 percent of its body oxygen in blood and muscle and only 5 percent in its lungs; humans, in contrast, hold about a quarter of their body oxygen in the lungs.

Emperor penguins, birds which can dive one-third of a mile deep, store almost half their body oxygen in muscle.

Carbon Monoxide and Other Poisons

The entire process of docking oxygen onto hemoglobin's iron sites is so delicately orchestrated that any interference with this process can be fatal. Carbon monoxide (CO) binds to the iron site in hemoglobin at least 250 times more strongly than oxygen does. As a result, when someone inhales carbon monoxide, it displaces the oxygen from the hemoglobin. People smother when carbon monoxide prevents half of their iron sites from accepting oxygen. Carbon monoxide normally occupies about 1 percent of the iron sites in healthy hemoglobin. Smokers who inhale may add 15 percent more carbon monoxide to their hemoglobin. Air in heavily trafficked, enclosed tunnels and parking garages and in homes with inefficient, hot unvented stoves often contains enough carbon monoxide to fill 16 percent of a hemoglobin's iron sites and cause headaches and shortness of breath.

The cyanides, poisons favored by many mystery writers, can kill within minutes. They interfere with the cellular uptake of oxygen by disrupting the function of cytochrome oxidase, the molecule responsible for 90 percent of the cell's uptake of molecular oxygen. The kernels of apples, peaches, and apricots contain cyanide or its immediate chemical precursors. Fruit seeds are not recommended as food, because they have made some people sick.

Rogue Hemoglobins

Nearly five hundred different forms of abnormal human hemoglobin exist. Happily, most mutant hemoglobins are chemically and medically harmless and quite rare. Most people with abnormal hemoglobins have no clinical symptoms and have been identified only when large populations were screened for hemoglobin types.

For example, only two or three families worldwide have the mutant called Rainier hemoglobin. Rainier binds oxygen thirty times more tightly than normal hemoglobin does, so oxygen loads on unusually easily at high altitudes or during strenuous exercise. However, this mutant hemoglobin does not release the oxygen easily. As if to compensate, the bone marrow makes more red blood cells than normal. Usually, individuals with the Rainier mutations function normally.

Even the smallest structural or chemical change can sometimes drastically affect hemoglobin's function and wreak havoc on the lives of its victims. Whether the alterations are induced by a genetic mutation or by a chemical accident, they can create dangerous, rogue hemoglobin.

What determines whether a hemoglobin mutation will be innocent or destructive? Recall that hemoglobin's chains form pockets to envelope the iron atoms and protect them from oxidation. Thus, disease occurs most often when a mutation or chemical change occurs inside the fold, near the heme, or on one of the amino acids.

Porphyrin Problems

European royalty have been plagued by an excess of princely porphyrins for at least five hundred years. George III, the English king who lost the American colonies, was only one of many members of Europe's royal families to suffer from an inherited porphyria disease.

The porphyrin is the ring of atoms that holds the iron atom in place inside a hemoglobin. If the iron atom cannot get to its place at the center of the ring, empty porphyrins enter the blood and the wayward iron accumulates in the liver.

Urine the color of red wine and bouts of intense abdominal pain followed by acute psychotic attacks mark this porphyrin disease. It has been traced in Europe's royal families as far back as Mary, Queen of Scots, in the sixteenth century. Her son, King James of England, described his urine as the color of his favorite port wine and said he inherited colic from his mother. Today, one in fifty thousand persons in Scotland suffers from the disease.

King George III of England, depicted in the movie *The Madness of King George,* suffered three short, but acute, attacks of abdominal pain followed by mania; his urine was also dark red. His physicians—and Parliament—thought he had gone mad. A metabolic disease was never imagined. Four of George's children—including Queen Victoria's father—may have inherited a kind of porphyria.

Several present-day descendants of George III have discolored urine and sun-sensitive skin.

During the eighteenth century, the founder of modern Germany, Frederick the Great of Prussia, was physically and psychologically abused by his father in explosive rages characteristic of acute intermittent porphyria.

In other types of porphyrin diseases, patients can be unusually sensitive to light. The porphyrin molecule is extremely photosensitive, and when it diffuses to the skin, it can cause severe scarring, spotting, and even disfigurement.

Hereditary factors are not the only cause of porphyrias. Toxic chemicals can also damage porphyrins: in 1959 several thousand Turks who ate seed grain treated with a fungicide became ill with a porphyria skin disease. Today even carriers of porphyria genes are warned that fasting, chronic liver disease, ethanol, barbiturates, estrogen, or certain noxious chemicals can precipitate an acute attack of porphyria.

Porphyrins can treat disease as well as cause them, though. A special porphyrin known as a hematoporphyrin is the basis for several cancer therapies in which the light-sensitive molecule is taken up by tumors and then stimulated with ultraviolet or laser light to release toxins that then kill the tumor cells.

Sickle Cell Disease and the Thalassemias

Sickle cell disease (formerly called sickle cell anemia) and the thalassemias are two of the most common—and devastating—genetic diseases in the world. Both are inherited, incurable blood disorders caused by abnormal hemoglobin synthesis. One small change in an amino acid makes the hemoglobin inside the red blood cell polymerize and cause sickle cell disease. A different problem causes the group of genetic blood diseases called the thalassemias. People with thalassemia can make normal hemoglobin chains, but they cannot make equal numbers of alpha and beta chains.

Sickle cell disease and the thalassemias occur wherever malaria is endemic, from parts of Africa, Italy, Greece, and Turkey through the mosquito-plagued swamps of the Middle East to India. Thalassemia's range extends through the Far East and Southeast Asia to Malaysia and Indonesia. Slavery brought sickle cell disease to North, Central, and South America and to the Caribbean.

Although they are caused by different genetic problems, both sickle cell disease and thalassemia are inherited in much the same way. To have the disease, a child must inherit two mutant genes. If both parents carry a mutant gene, each of their children stands a 25 percent chance of inheriting two mutant genes. If only one parent passes on a mutant gene, the child may have no symptoms; however, having one mutant gene imparts heightened immunity to the severe form of malaria. This immunity gives carriers a genetic advantage wherever malaria is endemic.

Children who inherit only one sickling gene usually survive their first bout with malaria and then acquire a natural immunity to the disease. In addition, the mutant hemoglobin actually helps kill the parasite that causes malaria. A particularly lethal form of malaria caused by the *Plasmodium falciparum* parasite is transmitted to people by the bite of the *Anopheles* mosquito. The parasites enter the red blood cells where, nourished by the globin, they multiply rapidly. But the red blood cells with defective hemoglobin give the parasite indigestion, so to speak, and the parasite fails to thrive. As a result, carriers of the gene have a critical advantage in malaria-infested parts of the world. In sub-Saharan Africa, the sickle cell gene can be found in 20 percent of the population.

In the past, many people who carried the sickle cell or thalassemia gene were reluctant to marry and have children for fear of passing on the disease. Today laboratory tests can be used to analyze the DNA in the placenta or in fetal skin cells obtained from the mother's amniotic fluid, and parents can decide whether to carry the pregnancy to term. Thus, genetic testing and counseling are important for carriers of this gene.

Worldwide, a quarter million babies are born with sickle cell disease each year, and one in twelve African Americans carries the sickle cell trait. The median age of death for the seventy-five thousand African Americans who actually have the disease is approximately forty-five years.

Sickle cell disease can cause shortness of breath, pain, strokes, blindness, arthritis, and kidney and heart failure and can damage the lungs, spleen, and hip joints. The damage caused by this disease is triggered by a tiny and seemingly insignificant change in one of hemoglobin's amino acids. Changing a glutamic acid to a valine at a sensitive spot on the beta chain transforms the mechanical properties of the red blood cells after they have delivered their oxygen to the tissues and organs. A normal red blood cell resembles a smooth, soft, and rather squishy doughnut. It is so flexible that it can squeeze through the capillaries even though they are much smaller in diameter. In people with sickle cell disease, however, the amino acid modification causes the deoxygenated hemoglobin molecule to form long stiff rods.

The trauma starts within a fraction of a second after the red blood cells discharge their oxygen. Suddenly, the hemoglobin molecules stick together and polymerize into long rods that rigidify the oxygen-free red blood cells and make them sickle shaped. As the sickled cells move through the capillaries, they create painful logjams that block the flow of blood and oxygen to the body (see figure 5.10). Sickled cells die faster than they can be replaced, and anemia occurs (hence the name sickle cell anemia).

The speed of polymerization (or gelation, as it is also called) depends on the amount of mutant hemoglobin in the red blood cell. In severe cases, polymerization can occur in as fast as ten-thousandths of a second. People with only one sickle gene produce so few mutant hemoglobin chains that polymerization is very slow. Scientists discovered that halving the amount of mutant hemoglobin in the cell increases the polymerization time to twenty-eight hours, giving deoxygenated blood cells plenty of time to squeeze through the capillaries and return to the lungs. As a result of this discovery, therapeutic drugs are being tested for their ability to lengthen polymerization times.

The presence of fetal hemoglobin also prevents the mutant hemoglobin from polymerizing. Mixing equal amounts of mutant and fetal hemoglobin increases polymerization times sevenfold, and the risk of early death declines as the amount of fetal hemoglobin in the blood increases. The presence of an unusually large amount of fetal hemoglobin has enabled many Bedouins with sickle cell disease to work in hot, dry Saudi Arabia under conditions that would normally have killed them. The Bedouins came from malarial areas of the Persian Gulf. Up to 20 percent of their hemoglobin was fetal, compared to about 8 percent in the average African American with sickle cell

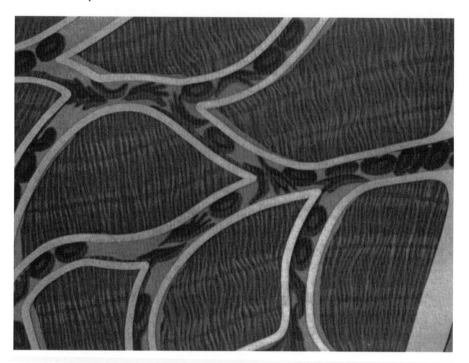

FIGURE 5.10. LOGJAMS PRODUCED BY SICKLED RED BLOOD CELLS.
The doughnuts in the diagram represent healthy red blood cells carrying oxygen. They are soft and squishy enough to squeeze through the capillaries. In sickle cell disease, some of them become long and stiff after releasing their oxygen. These sickled cells clog the capillaries and cause great pain. Reprinted with permission of the National Institutes of Health, National Heart, Lung and Blood Institute.

disease. The fetal hemoglobin took the place of adult hemoglobin in their blood cells. Bioengineers hope to create a genetic switch to turn on the production of fetal hemoglobin and, thus, prevent sickle cell disease.

Carriers of the sickle cell trait rarely experience life-threatening symptoms. However, sickling increases when oxygen pressure is low, when the concentration of mutant hemoglobin is high, and when temperatures soar. For example, when new recruits flew to a high-altitude United States Air Force base in Colorado and exercised strenuously during hot, dry weather, two recruits collapsed. They were sickle cell carriers, and even though their blood contained only small amounts of sickled hemoglobin, they died because the combination of heat, low oxygen pressure, and severe dehydration increased the concentration of mutant hemoglobin in their blood.

A completely different problem causes the genetic diseases called thalassemias. Healthy bone marrow makes equal numbers of alpha and beta chains, but in the thalassemias, too few alphas or betas are made, and some of the chains cannot find partners. Single alpha chains are toxic and poison blood cells. In a disease called hydrops fetalis, the most common genetic problem among southern Chinese, a baby is born with no alpha hemoglobin chains. The baby is stillborn or dies shortly after birth. Single beta chains join up in fours and form a normal-looking hemoglobin, but it has such a high affinity for oxygen that it cannot discharge enough oxygen into the tissues.

The genetic defect responsible for Cooley's anemia—which is also called beta thalassemia, thalassemia major, or Mediterranean anemia—occurs shortly before birth. At that time, as we have seen, the fetus normally stops making fetal gamma chains and boosts its production of beta chains. However, in Cooley's anemia, the beta gene does not switch on completely. Too few beta chains are made, and the unpaired alpha chains are highly toxic and poison red blood cells. Some fortunate patients can compensate for their faulty beta chain production by manufacturing enough extra gamma chains to pair off with the alphas as fetal hemoglobin. Armed with enough fetal hemoglobin, a person with thalassemia may be only mildly anemic.

Thalassemia and its treatment dramatize the danger of having too little—or too much—iron. A child diagnosed with Cooley's anemia before 1960 faced a devastating and disfiguring disease. In Cooley's anemia, the bone marrow tries to compensate for the body's lack of functional hemoglobin by expanding to fill every hollow in the bones, including the skull. As this expansion continues, bone walls thin and frequently break. Trying to help, the liver and spleen become grossly enlarged. Growth is retarded, cheekbones enlarge, the skull flattens, and there is malocclusion of the jaw. Untreated children acquire the chipmunk-like facial features that are characteristic of the disease. By the age of two or three, youngsters with untreated Cooley's anemia face congestive heart failure and, inevitably, death.

Starting about 1960, prospects for Cooley's anemia patients began to improve. Clinicians discovered that administering blood transfusions every two to four weeks countered the anemia and prevented damage to the bones and organs.

One of the tragedies of Cooley's anemia, however, is that both the disease and its treatment can kill. By the age of twelve, the typical patient has received more than 160 units of red blood cells. Of course these transfused cells contain iron, 0.25 grams of iron per unit, or a total of 40 grams. Humans normally function on 3 to 4 grams of iron. Because human bodies are designed to conserve iron and have no natural method for excreting it, almost all regularly transfused patients had accumulated toxic iron overloads by early adolescence. Free iron atoms react with oxygen and water in the body to form toxic free radicals that irreversibly damage the liver, heart, and pancreas. Teens developed liver disease, diabetes mellitus, and other complications and died slow and agonizing deaths from progressive heart failure. The blood transfusions that they needed to live were poisoning them. Unless the iron overload could be removed, death from iron toxicity was inevitable.

J. B. Neilands's 1950s discovery of the iron-grabbing siderophores in a soil fungus was the key to their survival (see chapter 3). Learning of Neilands's siderophores, chemists at a Swiss drug company—the Ciba-Geigy Pharmaceutical Company (now Novartis) in Basel, Switzerland—hunted for medically useful siderophores. In 1962, a naturally occurring siderophore, deferoxamine, sold under the trade name Desferal, came on the market. It is used to treat thalassemia and other anemias in which high iron absorption leads to toxic iron overloads. Clamped together, the siderophore and its iron atom pass through the body as a unit and are excreted in the urine or stool.

Clinicians struggled for almost thirty years before learning precisely how to administer Desferal in large enough quantities to remove all the excess iron from Cooley's anemia patients.

Since the early 1980s, Desferal has been delivered by pump, under the skin, for nine to twelve hours nightly. Enormous doses are needed: an intramuscular injection of a half gram of Desferal ejects only 5 milligrams of iron via the urine. The system is so time-consuming and cumbersome that many teenagers refuse to use it, with tragic consequences.

However, Cooley's anemia patients who continue this therapy have a 95 percent chance of surviving until they are twenty-five years old. Their growth and sexual maturation are normal or almost normal, and their vital organs are undamaged. Today, some live into their thirties and beyond. Many have fathered or given birth to children with normal, healthy hemoglobin. Without regular therapy, however, their odds of surviving beyond twenty-five drop to 30 percent.

As late as 1961, 15 percent of the population in parts of Italy and Greece carried the gene that causes Cooley's anemia. In southern Sardinia, 30 percent of the people had it. An estimated 100,000 to 300,000 infants are born yearly with severe forms of thalassemias. At least 6.5 percent of children worldwide carry a thalassemia gene.

Because of the high incidence of the disease and the enormous cost of treatment in Greece, the army screened the population for thalassemia genes during the 1970s. Pregnant women who had the trait and whose mates also had the trait were offered prenatal diagnosis and selective abortion. Both Roman Catholic and Greek Orthodox churches looked the other way because the cost of caring for large numbers of Cooley's anemia patients would have swamped the country's medical system. As a result, the number of severe thalassemia cases born in Greece dropped to 10 percent of previous levels. In Sardinia and Cyprus, almost no children are now born with either sickle cell disease or Cooley's anemia. In mainland Italy, birth rates of children with hemoglobin diseases have dropped to 20 percent of previous projections. In the United States, the mean age of patients with Cooley's anemia is 16.2 years.

So far, thalassemia is incurable except in a few patients who have a brother or sister with the same tissue type. In these cases, bone marrow transplants are sometimes effective.

Scientists hope to develop an oral chelation therapy and a way to increase the supply of fetal hemoglobin in patients with severe thalassemia. Ultimately, gene therapy may be used to transfer a functioning beta- or alpha-chain gene into a patient's bone marrow.

Artificial Hemoglobin

An artificial substitute for hemoglobin is greatly needed. Whenever there is severe bleeding and 40 percent of the body's blood is lost, transfusions are needed to restore the oxygen-transporting functions of the lost hemoglobin. Many surgical procedures, for example, cannot take place without an adequate supply of blood for emergency transfusions.

Blood supplies are not expanding as fast as the world's population, and the part of the population that uses the most blood, namely, the elderly, is increasing at even faster rates than the general population. Experts predict that by the year 2030, the United States alone will face an annual shortage of 4 million units of blood. A unit of blood is 500 milliliters.

Besides the world's growing demand for blood, current supply systems face other problems too. Because of the immune properties of blood, an error made in transfusing one type of blood into someone with another blood type can be extremely serious. Yet the shelf life of blood is only about two months, so the blood of a particular type may not be available when and where it is needed.

Blood products can also transmit infections. Although blood is normally tested carefully for known pathogens, the fear of infection has decreased the public's willingness to donate blood. In the United States, only 5 percent of the population donates blood.

Given hemoglobin's size and complexity, it is not surprising that attempts to replace it in blood have proven difficult. Most efforts have concentrated on replacing hemoglobin's oxygen-transporting functions. Synthesizing hemoglobin has proved impractical, however, because the molecule is so big and complex and because pure hemoglobin does not function normally when it is repackaged in red blood cells. Another research approach has focused on using perfluorocarbons to increase the solubility of oxygen in blood so that blood can carry more oxygen. However, the chemicals tested so far have toxicity problems and do not deliver enough oxygen to the tissues.

Recycling the hemoglobin in old donated blood is another possibility. Hemoglobin is taken out of the red blood cells, sterilized, and stabilized. This "cell-free" hemoglobin can be stored for long periods of time. It can be given to anyone, regardless of blood type, and it greatly reduces the chance of infection. It also transports oxygen fairly well. Unfortunately, it does not stay in the bloodstream long enough, and it has some undesirable side effects. Like most blood transfused today, it depends on the willingness of donors to give blood.

Ultimately, genetically engineering human hemoglobin may solve the impending blood crisis. The gene for human hemoglobin has been isolated. Inserted into bacteria, the gene can make the bacteria produce human hemoglobin. Methods for encapsulating free hemoglobin inside synthetic cells are under development. These technologies provide hope that a serious blood shortage can be averted. They take time, however, and development costs are expected to make blood substitutes several times more expensive than donated blood. So, in the meantime . . . donate blood.

Hemoglobin, this huge and convoluted molecule built for transporting oxygen, is one of the wonders of the natural world, but it is not the only iron-based system that vertebrates developed over time. Amazingly enough, some birds, fish, sea turtles, and other animals have evolved migration patterns by making and using their own tiny iron magnets that sense the magnetic field created by Earth's iron core.

While hemoglobin and myoglobin constitute a small and quite well understood system, magnetic migration is geographically sprawling, and it involves a magnetic detector that we humans do not possess. Thus, some of the world's most spectacular travelers detect Earth's magnetism and probably assemble iron-based detectors like the magnets manufactured by oxygen-leery bacteria. Until recently, this molecular guidance system was not intensively studied. So only now are we beginning to learn about these magnificent vertebrates and their iron-based, magnetic migratory skills.

6

Migrating Animals
MAGNETIC TRAVEL

Of the billions of creatures that migrate across our planet every year, the Arctic tern holds the records. Each year, this dainty bird—not much bigger than a robin—flies more than 22,000 miles, traveling almost pole to pole and back again. In a lifetime of thirty-four years, one banded tern flew more than 600,000 miles. At least ten species of birds commute between the polar regions. Dozens of others journey thousands of miles between the Northern and Southern Hemispheres. Scores more perform lesser feats, traveling hundreds of miles between summer nesting regions and winter feeding grounds. Many of these migrants fly at night; all must contend with cloudy skies and fog. In some species, untutored fledglings barely out of their nests make their first migration alone, unaccompanied by older and experienced birds.

How can any species count on successfully completing such long and hazardous trips year after year after year? How can we learn what birds, whales, salmon, sea turtles, and other long-distance travelers need to survive their journeys? How can scientists conduct the exquisitely controlled experiments needed to answer these questions? After all, many migrating animals are too big, too nutritionally demanding, or too far-flung and long-lived to be carefully observed in the laboratory.

Precisely because they are small and manageable, birds have become the most extensively studied navigators in nature. A homing pigeon released 400 miles away finds its way unaided back to its laboratory cubbyhole. Many migrating birds also follow predictable paths, flying in a more or less constant direction over several weeks to reach feeding grounds. In both hemispheres, birds travel well-known highways. In North America, many find their way across the Great Lakes via two opposing peninsulas that jut into Lake Erie. In Europe, thousands of birds cross a similar "bridge" at Gibraltar. "Birds of a feather flock together," often on the same tight schedule, so that all arrive at their summer homes within a week or two of one another. There they often settle into the very tree, bush, or tuft of grass where they nested the previous year. Spring migrations, when birds are driven to breed and raise young, are particularly frenzied. So great is the urge to travel that migratory songbirds caged in laboratories become restless even in autumn, frantically hopping up and down, facing the same direction that their free-flying comrades are racing toward. Biologists have seized upon this restlessness, manipulating it to reveal nature's migratory secrets.

Birds fly equipped with a veritable toolbox of navigational skills, redundancies, and backup systems. The sun and stars help them in fair weather, but during the 1980s biologists realized that birds also use planetary forces that human beings cannot sense. When skies are dark or overcast and celestial cues obscured, some bird species turn to one of Earth's characteristic forces to navigate the globe. Like magnetic bacteria, these birds carry their own magnetic compasses and use the planet's magnetic field to guide them on their way.

In 1963, a small thrush-like songbird called the European robin, *Erithacus rubecula* (family Turdidae), first demonstrated avian use of magnetic orientation, but years passed before the biology community embraced the concept. Unlike magnetic bacteria, which were immediately accepted, the discovery of magnetic orientation in larger creatures was plagued by pitfalls and pratfalls, by complexities and misunderstandings, and by years of deep skepticism and outright rejection.

For centuries, nothing had been known about bird navigation. A suggestion made in 1859 that animals might sense Earth's magnetic field was ignored. In the mid–twentieth century, the hypothesis was raised again when pigeon racers and the United States Army Signal Corps noticed that pigeons veered from their routes during magnetic storms, which coincide with sunspot activity. Logs kept by the Italian Pigeon Lovers Federation of Parma on twelve thousand pigeons raced between 1932 and 1957 confirmed these observations. However, the first attempt, in the 1940s, to test the hypothesis was unsuccessful. Then, during the 1950s, it was discovered that birds can steer by solar and stellar compasses.

Swept up in the excitement about these seemingly exotic modes of orientation, a few researchers revived the idea that birds could also use Earth's magnetic field to navigate. Not all planets have magnetic fields, but ours does because it has an iron core. At its center is a solid iron mass almost as large as the moon. Around this solid core is a layer of hot liquid iron that rotates once every 120 years or more. The positive and negative charges of this circulating molten iron constitute an electrical current, which, like all electrical currents, generates a magnetic field.

Intrigued, this little band of visionary scientists concluded—with more faith than evidence—that birds could sense magnetic fields much the way humans unconsciously sense Earth's gravitational pull to intuit up from down. Although time would prove these researchers correct, they had no evidence to document their hunch. Today it is generally agreed that they were suffering from "overexcitement," and their claims cast a pall over the hypothesis of magnetic orientation for decades.

Thanks largely to the persistence of Professor Friedrich Merkel and his colleagues and successors at the University of Frankfurt in Germany, magnetic orientation was accepted back into the fold of scientific respectability in the 1980s, thirty years later. Merkel was studying how the European robin uses star patterns to orient at night. The European robin, Britain's national bird, is smaller than the American robin but has a similar rusty front. The European robin summers throughout Europe in forests and suburban gardens. Each autumn, some migrate to southern Europe while others cross the Mediterranean at Gibraltar to winter in North Africa.

During the autumn migration season, the European robins in Merkel's laboratory behaved normally enough during the day. But European robins are nocturnal travelers, and at night they became restless in their cages. Almost hyperactive, they fluttered their wings and hopped up and down on their perches on the side of their cages closest to the direction of their would-be migration. Germans called this behavior *Zugunruhe,* migratory unrest, a term that scientists worldwide still use.

One of Merkel's graduate students, Hans Gerhard Fromme, designed a cage for testing the robins. It was octagonal with eight perches radiating from the center like the hand of a clock pointing to 12:00,

1:30, 3:00, and so on. On the cage floor Fromme placed a sheet of paper coated with typists' white correction ink. Wherever a bird hopped from the perch onto the floor it left a scratch on the paper. Counting the scratches, Fromme could determine the direction the birds faced most frequently.

One evening during the autumn migration season, Fromme tested some robins on the roof of a university building. To prevent the birds from using the stars or city lights for orientation, he shielded their cage with translucent plastic. Checking the next morning, Fromme was surprised to find an unmistakable pattern of scratches on the cage floor. Although the birds could not have gotten their bearings from stars or city lights, they had still pointed southwest—toward Gibraltar. Merkel's group deduced that something must be giving them directions when the stars are obscured.

During the early 1960s, when Fromme was working on his Ph.D. thesis, radioastronomy was a new and thrilling field of physics. Two new classes of heavenly objects—quasars and pulsars—had been discovered emitting enormous amounts of radio wave energy, mostly at long wavelengths. A physicist would have realized that the antenna needed to detect such long waves would be much too large for any bird to carry. But Merkel's group, not composed of physicists and astronomers, thought the birds could sense these long wavelengths. So one of Merkel's graduate students, Wolfgang Wiltschko, currently a professor at the University of Frankfurt, worked in happy ignorance designing an elaborate experiment to prove the group's hypothesis. Waiting for his oh-so-clever equipment to be built, Wiltschko spent the time testing the robins in the laboratory's steel hypobaric chamber. These spheres—literally, low-pressure chambers—are routinely used to simulate the low-oxygen conditions experienced by high-altitude mountain climbers and astronauts. Inside the windowless chamber, the robins could not possibly see the sun or stars. Sure enough, when placed inside the sphere, the robins became disoriented and hopped and fluttered randomly every which way. The birds were surely not receiving any orientation cues from pulsars, quasars, stars, or sun. Fromme and Merkel suspected that the magnetic field might be distorted inside the chamber. So after putting a bird back inside the chamber, Wiltschko turned on a weak magnetic field to mimic Frankfurt's natural magnetic conditions.

"And this was done on the 12th of October of 1963," Wiltschko burst out in a rush of enthusiasm. "That is the date of the discovery of magnetic orientation." During the autumn migration, when the bird would naturally head southwest, the robin named R-2 had hopped more than a thousand times onto its southwestern-most perch.

Yet Wiltschko could not believe it. "It was against all my theories and all my prepared ideas." So the next day he tested all his robins again and again. No matter what the direction of the field, the robins always oriented themselves toward the magnetic coil's southwest. Without any celestial cues at all, the European robin was orienting itself toward Gibraltar. In short, the bird was using a magnetic compass.

"The funny thing for me still is that I discovered magnetic orientation with a totally wrong hypothesis," Wiltschko laughed. What if all his clever equipment for his "ingenious" test of quasars and pulsars had been built sooner? "I would still have discovered it, but it would have been later," he declared confidently.

Controversy

Thirty years later, Wiltschko's experiment is regarded as "key" and "compelling." When it was published in 1968, however, his report was greeted with skepticism. Scientists in other laboratories, using different cage designs and computing their averages differently, could not duplicate the experiment. Most important, the results of the experiment were not visually spectacular. Magnetic effects are seldom eye-catching. As Kenneth Able, a professor of biology at State University of New York in Albany and a leading American researcher in magnetic orientation, has observed, "There were many, many hops along all the perches, and the differences between them was really slight. These were legitimate concerns at the time, but it's sort of the way science works. If people are disinclined to believe something when it first comes out, it's amazing how critical they can be about finding flaws in the experimental design, the statistics, and this, and that. But if you come out with a result that people are predisposed to believe, the level of criticism is remarkably reduced."

There was also no explanation of how birds might sense magnetic fields. Particles of magnetite had not yet been found in birds, bacteria, or any other living system. And there were still memories of those early unsuccessful attempts to prove magnetic orientation in the 1950s. Although NASA was intrigued enough with exotic new modes of navigation to invite Wiltschko to explain his work at a 1970 conference, scientists thought his reports farfetched.

Meanwhile, Wiltschko, assigned to teach a class on fish, had fallen in love with a cheerful young student, Roswitha Brill. When Wolf and Rosie married, she switched from fish to magnetic orientation. She knew from the beginning that the field was controversial. "Many scientists hesitated to accept the idea that animals might make widespread use of a factor which man is so obviously unable to detect," she said. Kenneth Able concurs, "There was a climate of doubt about the whole idea. Even to work on it or to talk about it was like parapsychology, not something you want to do."

In 1972, Rosie and Wolf Wiltschko dropped another bombshell in the first article they wrote together as postdoctoral fellows. They had discovered that although European robins use Earth's magnetic field to migrate, they do not use it in the same way that a hiker uses a conventional compass to navigate through the forest.

Any traveler, beast or human, in parts unknown needs both a map and a compass to find the way home. One is useless without the other. Bushwhackers, for example, need to know where they are relative to their goal. Looking at a map, they can see that to go from Florida to New York they must travel north. But to locate north, they need a compass. Ancient navigators used star charts for maps and magnetic lodestone needles for compasses. A homing or migrating animal needs much the same equipment because, as it migrates, it must continually calculate both its location and its destination.

The familiar hiking compass has a thin magnetic needle that rotates horizontally to point to the North pole; the end of the needle that points North is often marked with red paint. If the compass needle is centered on a well-engineered bearing, it also dips up or down at an angle that depends on the inclination of the magnetic field at the compass's location (see figure 6.1 and chapter 4). Like magnetic bacteria, the needle orients along the direction of Earth's magnetic field lines.

In the Southern Hemisphere, the red tip of the compass needle points up away from Earth, parallel with the magnetic field; and in the

Northern Hemisphere, the red end points down toward Earth, also parallel with the magnetic field. At the magnetic equator, the needle is flat, parallel with Earth's surface and its magnetic field. But because the compass needle holds itself parallel to Earth's field, in the Northern Hemisphere its red end dips down more steeply the closer it moves to the Arctic. In the Southern Hemisphere, it points up more sharply as it moves closer to Antarctica. At the magnetic poles, the compass needle points either straight up or straight down. What kind of a compass would a European robin have?

To find out, the Wiltschkos designed a new experiment. By now, advancements in technology allowed them to use computers: whenever a bird hopped on its perch, it activated an electronic switch that was recorded on paper tape and analyzed by a computer in the Wiltschko's laboratory. An electromagnetic coil around the birds' cages enveloped the birds in an artificial magnetic field, which the Wiltschkos could systematically vary without harming the birds. They kept some European robins in Frankfurt's natural magnetic field. For other birds, they changed not just the north-south direction of the magnetic field lines but also the compass needle's dip. When they did so, they discovered that European robins do not respond to changes in the north-south axis of the field lines. Instead, the birds were acutely sensitive to their dip: that is, how steeply the magnetic field lines incline in or out of Earth. As a result, the robins do not orient northward or southward. Instead, they move poleward or equatorward; that is, they use an inclination compass. Since the discovery of the dip compass in robins, the phenomenon has been reported in seventeen other bird species.

Why would a migrating robin want a dip compass instead of a polar-pointing compass? The answer is that because a dip compass is sensitive to the inclination of Earth's magnetic field, it can be used to determine north-south position, that is, latitude; and thus it can help any species that commutes between north and south. Savannah sparrows, *Passerculus sandwichensis,* for example, can nest anywhere between 40 and 60 degrees north latitude. In autumn as they fly toward the equator and again in spring on their flight toward the north, their dip compasses sense changes in the inclination of Earth's field and tell the sparrows where they are.

The fact that many bird species orient with dip compasses is widely

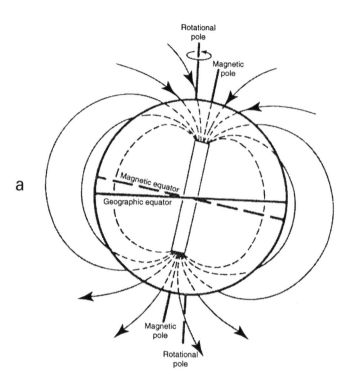

a

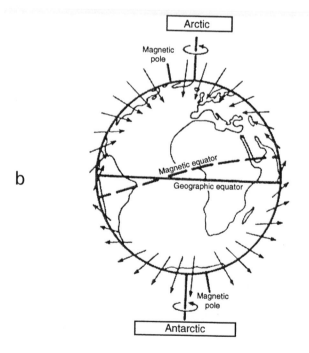

b

accepted today, but at the time it was dismissed out-of-hand as outlandish and unbelievable. And the Wiltschkos' dip compass experiment—considered "elegant" and "classic" today—just intensified the scientific community's doubts about magnetic orientation.

Homing Pigeons

A species of nonmigrators, homing pigeons, became the key to demonstrating the existence of magnetic orientation in migrating birds. Homing pigeons, descendants of the Mediterranean rock dove, *Columba livia,* are only slightly larger than their city pigeon cousins. Taken hundreds or thousands of miles from their nests and released, homing pigeons need only a few minutes to figure out the general direction of home. Then they race back—by quite direct routes—at speeds up to 50 miles per hour.

Their homing instinct has been well known for several thousand years. Ancient Egyptian soldiers released homing pigeons at the borders of their kingdom to inform headquarters about enemy troop movements. An athlete in ancient Greece released a homing pigeon to tell kinsmen of his victory in the Olympic Games. During the 1940s, the United States Army Signal Corps supported research on homing pigeons, and today Colorado River rafting guides dispatch carrier pigeons so that tourists' film can be developed by trip's end. During Prohibition in the 1920s, a New York gangster communicated via carrier pigeon with smugglers' ships laden with Scotch whiskey for the famous Cotton Club. As recently as 1999, an extortionist threatened to poison grocery store products in Germany unless he received $20 million in diamonds. He supplied a carrier pigeon with a backpack to fly

FIGURE 6.1. EARTH'S MAGNETIC FIELD.
(a) *Earth's magnetic field is like the field of a giant bar magnet. The force lines of Earth's geomagnetic field emerge from the earth in Antarctica and reenter Earth in the Arctic.* (b) *The arrows at Earth's surface show the direction a compass needle would point. Note that at the magnetic equator the compass needle is flat, parallel to Earth's surface. In contrast, a needle at the magnetic poles is perpendicular to Earth's surface. From R. Wiltschko and W. Wiltschko,* Magnetic Orientation in Animals *(Berlin: Springer-Verlag, 1995). Reprinted with permission.*

the diamonds to him. His nefarious plot was foiled, no doubt, by a stool pigeon.

Oddly enough, homing pigeons perform their prodigious flying feats not because they are migratory but because they are basically stay-at-homes. In a pigeon loft, each bird occupies its cubbyhole for life, hence the term "pigeonhole." Males help their mates care for the young, and centuries of breeding have sharpened this domestic drive. Hobbyists stage races during the pigeons' breeding season, taking males from the nest knowing that the birds will fly back as fast as possible to care for their young. (Scientists frown on this practice and fly their pigeons year-round.) Racing enthusiasts have been known to pay $180,000 for a champion homer.

Because of the pigeons' extraordinary desire to stay at home, biologists use them to study magnetic orientation during flight. Homing pigeons are the only nonmigratory birds whose magnetic orientation has been studied. They are also among the few birds outside the large category of passerine, or perching, birds whose use of magnetic fields has been analyzed.

In 1971, William T. Keeton, a biologist at Cornell University, decided to see what would happen if he interfered with the homing pigeons' natural magnetic environment. After gluing bar magnets to the backs of his pigeons' necks, he released the birds in both sunny and cloudy weather. A group of control pigeons carrying brass bars of the same size and weight were released at the same time. Not being magnetic, of course, the brass bars would not disturb the birds' magnetic surroundings. On sunny days, both groups flew straight home as usual. But on totally overcast days, with the sun nowhere to be seen, the birds with brass bars made it home, whereas those carrying the disruptive magnets flew off in random directions. Keeton concluded that when the sun is obscured, pigeons switch from their then-useless sun compass to a magnetic compass.

Keeton's experiment with pigeons revealed little about magnetic migration that the Wiltschkos had not already demonstrated with their European robins. Although Keeton had used magnetic fields to disturb the birds' homing ability, he had not demonstrated unequivocally that changing the magnetic field made them change direction. Nevertheless, Keeton was highly respected in the homing pigeon field, and biologists began to pay attention.

Keeton invited the Wiltschkos to Cornell in upstate New York and taught them how to raise and keep homing pigeons. Today the Wiltschkos tend up to five hundred pigeons in downtown Frankfurt, next to a city park. Some of the pigeons live on the roof of the university's biology building, while others are sheltered under trees in a bedroom-size cage fronted with wire mesh. The inside wall of the cage has square cubbyholes painted in primary colors with a number above each nest. The dovecote looks altogether like a kindergarten with cubbyholes for mittens and toys, except that everything from chest-height down is covered with white bird droppings and, in molting season, feathers. Outside the dovecote, a flock of pigeons wheels and swoops up and down and around over the park like a school of fish. Homing pigeons generally live six or seven years, but the Wiltschko employed one bird for seventeen years. Thanks to Keeton's help, the Wiltschkos could conduct behavioral experiments with a group of birds that would always return home, bringing their tracking equipment with them.

The ability of homing pigeons to actually sense and respond to Earth's magnetic field was finally tested by Charles Walcott and Robert Green, biologists at State University of New York at Stony Brook on Long Island. In 1974, they collared and crowned pigeons with miniaturized, battery-powered magnetic coils. Each bird also carried a tiny radio transmitter inside a bird-size backpack to send the bird's bearings back to Walcott and Green on the ground (see figure 6.2). Dividing the pigeons into two groups, Walcott and Green gave each group a diametrically opposite magnetic field. In the first group, the dip compasses pointed up, and in the second they pointed down. On sunny days, both groups relied on visual cues from the sun and flew directly home. But Walcott and Green predicted that on solidly overcast days, the birds would switch to magnetic cues and fly in opposite directions. And sure enough, they did. Clearly, the pigeons used a compass to orient, just as the Wiltschkos' European robins had done. As Kenneth Able put it, *"That's* the experiment that provides strong evidence for a magnetic compass in homing pigeons because you get those pigeons, you predict which way they're going to go, and they go that way. That's a really strong result. It's a wonderful experiment." When an Italian laboratory replicated Walcott and Green's experiment, biologists really began to sit up and take notice.

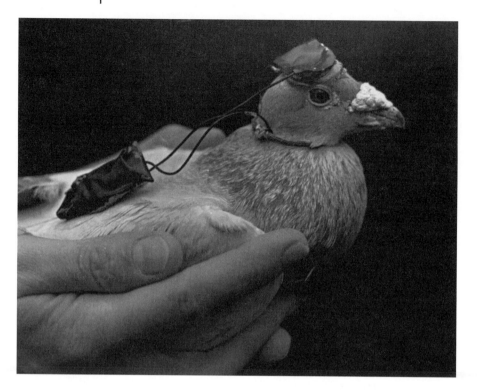

FIGURE 6.2. A HOMING PIGEON WITH ITS BATTERY PACK.
Reprinted with permission of Charles Walcott.

Today, the fight about magnetic orientation is largely over. "There's just the accumulation of so many data from so many different kinds of organisms and from so many different laboratories," Able observed. "The essence of science is replication, and there was enough replication of these results that finally the weight of evidence was overwhelming. There was too much to ignore." James Gould of Princeton University phrased it more cynically, "The people who thought that there was a problem (with the phenomenon) have all died or retired by now. That's the way a lot of things happen in the field, I'm afraid." On the other hand, as Gould and Able also emphasized, skepticism eventually forced biologists to produce better experiments, clearer data, and more believable work.

Only a few remain unconvinced. Oddly enough, one of them is

Donald Griffin, the discoverer of bat echolocation. Like magnetic orientation, echolocation is a sensory mechanism, an information-gathering system that humans are blind to. When Griffin searched for magnetic orientation in birds during the 1930s, he could not find it. "It's a puzzle," Griffin explained in 1999. "These are all very small effects requiring statistics to distinguish them from the background noise. It's odd that it's such an elusive thing." But even Griffin conceded in 1999, "I don't argue that there isn't orientation."

Ironically, the Wiltschkos themselves have gone through similar periods of disbelief about other phenomena. An Italian group of biologists led by Floriano Papi contended that pigeons use smell to help them home. After years of skepticism, Wolf Wiltschko finally visited the Italian researchers and discovered that their dovecote stood on a hillside, open on all sides to breezes. Back in Frankfurt, the Wiltschkos had used a Cornell-style cage, enclosed on three sides, but when they built an open-air Italian-style cage on the roof of their Frankfurt building, their pigeons also used scent to guide them back. The controversy took years to sort out.

Songbirds on Virtual Reality Journeys

Despite all this excitement, no one had demonstrated yet that birds actually sensed changes in magnetic fields along their migration routes and altered course as a result. So the Wiltschkos and graduate student Willi Beck decided to send a group of small songbirds called pied flycatchers, *Ficedula hypoleuca,* on a virtual reality journey. Pied flycatchers summer in Europe and winter in North Africa, and each autumn, Frankfurt's pied flycatchers head southwest toward the Strait of Gibraltar to cross the Mediterranean. There they turn left to fly southeast into North Africa for the winter.

Surrounding cages of pied flycatchers with magnetic fields, Beck mimicked the field intensities and inclinations that the birds would have met each day if they had flown freely along their autumn migration route. Early in their virtual reality trip, the birds oriented themselves toward the southwest and Spain. Later in the season, when the field intensities and inclinations in the laboratory imitated those of the Iberian Peninsula, the birds veered left toward North Africa.

Given the magnetic field data for their journey, the birds were genetically coded to follow it as instinctively as they breathed.

Meanwhile, a control group of caged pied flycatchers was enveloped in the natural magnetic field of Frankfurt; and these birds also hopped in the direction of Spain early in the migration season. But at the time they would have been arriving at Gibraltar they were still in Frankfurt's magnetic field. And as the free-flying pied flycatchers arrived in North Africa, the control birds in Frankfurt were hopping randomly and confusedly around their cages. They had not received their normal magnetic migration cues to turn left.

Pied flycatchers do not cross the magnetic equator during their migration, but another common songbird, the European garden warbler, *Sylvia borin,* crosses it twice each year. Germany's population of garden warblers summers in parks and woodlands and winters south of the equator in sub-Saharan Africa. What do garden warblers do when they reach the magnetic equator? Garden warblers also have dip compasses that sense whether Earth's magnetic lines of force point down into the planet's core or emerge up from it. At the equator, however, there is no dip to a compass needle because the lines of force are flat, parallel with the globe's surface. Does the garden warbler's dip compass give the bird contradictory directions at the equator, or no directions at all? Experiments with a transequatorial species such as the garden warbler were clearly called for, and the Wiltschkos carried them out.

During August and September of 1992, as a control group of wild garden warblers flew out of Frankfurt headed south, the Wiltschkos' caged garden warblers oriented south too. In October, as the free-flying warblers crossed the equator, the Wiltschkos surrounded the caged birds for two days with a magnetic field that mimicked the equator's magnetic field, a flat field without inclination. Then the equatorial field was switched off, and the birds were switched back to Frankfurt's field. At that point, something remarkable occurred: the caged birds turned around and began hopping in the opposite direction. Why had they reversed themselves?

While traveling in a virtual Northern Hemisphere, the caged warblers had navigated using the fact that Earth's field lines point down in that hemisphere. At the virtual equator, a neurological trigger notified the birds that they must follow the dip the opposite way. Once in the

Southern Hemisphere, they could continue south to reach their feeding grounds in South Africa only if they reset their dip compass to operate on field lines emerging up from Earth. When biologist Robert C. Beason of State University of New York at Geneseo duplicated the experiment with another long-distance migrant, the American bobolink, *Dolichonyx oryzivorus,* he got the same remarkable results.

Late one night in a bar, Wolf Wiltschko confided to a colleague that he had started the garden warbler experiment in 1992 already knowing what the birds would do at the equator. He revealed that while writing his Ph.D. thesis back in the 1960s, he had conducted a similar experiment but dropped it, sure that he could not get it published. The biological world was not yet ready to accept such a "far out" phenomenon, especially not one observed by a mere graduate student.

As more scientists became interested in magnetic orientation, it became clear that the phenomenon is probably quite widespread among migratory birds. Four separate research groups in the United States, Germany, Italy, and Australia have demonstrated that species representing quite different evolutionary lineages use magnetic fields to direct their flights. Dip compasses have been discovered in every species studied, including the American bobolink; two European warblers, the blackcap, *Sylvia atricapilla,* and the goldcrest, *Regulus regulus;* and two Australian species on the other side of the equator, the yellow-faced honeyeater, *Lichenostomus chrysops,* and the silvereye, *Zosterops lateralis.* Together, scientists from a variety of disciplines are assembling a remarkable picture explaining how birds migrate around our planet with such success.

According to one popular hypothesis, young birds who have not yet mastered the visual cues provided by the sun, stars, and landscape use magnetic orientation during their first flights. Some two- or three-month-old songbirds perform their first migration without any older experienced individuals along for guidance; yet these young birds can chart a general direction and get within their wintering range. Even hand-raised laboratory birds that have never seen the sun or stars can orient by the ambient magnetic field of the laboratory and its cycle of light and dark. Young pigeons are thought to record the direction of their outward journeys with magnetic orientation, and then reverse their compasses for a home course. Older pigeons use their magnetic sense to get their homeward bearings but once near home identify their specific landing site visually. For example, mature pigeons supplied with frosted contact lenses have no difficulty getting an initial home bearing but run into trouble a few miles from home. Apparently they need eyesight to locate their loft.

Most migrating birds on our planet travel between north and south as their food ripens. The typical bird travels a day and then spends the next several days resting, eating, and fattening up before flying off again. Birds receive magnetic field stimuli over a vast range of intensities. At any one time, however, a bird can sense only a rather narrow range of magnetic field intensities corresponding to whatever field the bird has grown accustomed to. Experiencing a field that deviates more than 25 or 30 percent from that can be disorienting. Fortunately, grown Savannah sparrows need only four clear days and nights to recalibrate their compasses and adapt to new ambient magnetic conditions. Ken Able and his research partner and wife, Mary A. Able, of State University of New York at Albany, have shown that Savannah sparrows can recalibrate their compasses throughout adulthood. The Ables' findings contradict the conventional scientific wisdom that animals can learn particular skills only during certain critical periods, generally early in life.

Since it can take a bird four days to recalibrate its compass, unusual magnetic occurrences along a migratory route can disrupt a bird's orientation. In Sweden, a particular 7-mile-long region has a magnetic field that is much stronger than that of the surrounding areas, probably because of local deposits of iron-rich ores. When migrating birds reach it, both night and day fliers quickly drop in altitude.

Colored lights can also interfere with a bird's ability to orient magnetically. Red light temporarily destroys the magnetic orientation of Australian silvereyes, and red and yellow lights disorient European robins, a puzzling phenomenon that scientists do not fully understand.

Many mature birds receive directional data from a variety of sensory systems and switch back and forth between them as conditions change. For nocturnal migrants, sunset is an important time because they decide then whether to fly that particular night and what direction to take. On clear evenings, migrating blackcaps and European robins begin their flights at dusk by studying cues from polarized light produced as the sun sets and the horizon glows; later, the birds study star patterns in the night sky. They may also use permanent landscape features. Close to their goal, some birds recognize local smells and familiar sights. Internal factors direct birds too: those that have grown rich in body fat may try to cross the Alps, leaving leaner kin to detour.

Nevertheless, on overcast days and cloudy nights when celestial cues are obscured and unavailable, many migratory birds can fall back on magnetic orientation, their most stable and fundamental navigational tool, their ultimate backup system, and one of our planet's primary features. Abandoned by sight and smell, they turn to and navigate by Earth's magnetic field. Several million years of evolution have provided for the inheritance of genetic characteristics advantageous to survival.

Sea Turtles

As evidence mounted that many migratory bird species steer by magnetic fields, scientists became curious about the orienteering skills of other animals. Sea turtles soon attracted attention because these large and powerful animals swim continuously for weeks through seemingly featureless waters with the dead-reckoning skills of an old-time sea captain. Had sea turtles also developed the ability to sense Earth's magnetic field?

Seconds after digging clear of their underground nests in Florida, a frenzy of 2-inch-long sea turtles scampers across the beach toward the dimly lit horizon, dives under the waves, and starts swimming hundreds of miles across the open Atlantic. Loggerhead sea turtles, *Caretta caretta,* spend their lives in almost continuous migration. After five to ten years in the open ocean feeding in Mediterranean currents, most of these turtles take up residence in feeding grounds along the North Atlantic coast. When fully mature in twenty-odd years, female loggerheads migrate back to lay their eggs on the same Florida beach where they were born.

Other species of sea turtles perform similarly remarkable feats of orienteering. The loggerhead turtles of the Pacific Ocean emerge from nests in Japan and traverse the Pacific to feed in Baja California. Kemp's ridley turtles, *Lepidochalys kempi,* converge from different points in the Atlantic, Caribbean, and Gulf of Mexico to lay their eggs on the same Mexican beach. Green turtles, *Chelonia mydas,* commute more than 1,200 miles each way between Ascension Island in the South Atlantic and their Brazilian feeding grounds. These turtles know precisely where they are going. Satellites have tracked green turtles swimming toward their destinations in direct, straight lines despite darkness, fog, stormy weather, a total lack of visual landmarks, and such poor eyesight above water that they cannot possibly be using star maps.

Every spot on the globe, including the turtles' nesting beaches and feeding regions, is characterized by two magnetic properties: the intensity of Earth's magnetic field and the angular dip of the compass (magnetic inclination). A husband and wife team of biologists, Kenneth J. and Catherine M. F. Lohmann at the University of North Carolina, have designed experiments which study the magnetic response of loggerhead turtles swimming in the North Atlantic. After leaving their natal beaches of the southeastern United States, the turtles enter the North Atlantic gyre, a circular current enclosing the food-rich Sargasso Sea (see figure 6.3). Only by staying safely within the confines of this warm-water nursery can the turtle babies find enough food to eat and grow into adulthood. After the turtles drift north in the gyre along the coast of the United States, they reach a point where they must turn eastward and cross the Atlantic to avoid the cold, killing waters that lie north of the Sargasso Sea. On the European side of the Atlantic the turtles reach a second point, near Portugal, where they must turn again to stay within the circle.

The Lohmanns tested the ability of the turtles to respond to Earth's magnetic field at these two turning points. The Lohmanns dressed the hatchlings in turquoise Lycra vests and attached the spandex to electronic sensors and computers (see figure 6.4). Then they put the tiny vested turtles in a swimming pool not much different from a child's wading pool. Hatchlings exposed to the magnetic intensity found at the American side (point 1, figure 6.3) and the European side (point 2, figure 6.3) oriented themselves so that they would remain in the safety of the gyre. Separate experiments showed that these turtles' compass also measured magnetic inclination. Clearly, loggerhead sea turtles could use Earth's magnetic field to assess their global position.

FIGURE 6.3. THE NORTH ATLANTIC GYRE.
From K. J. Lohmann and C.M.F. Lohmann, "Detection of Magnetic Field Intensity by Sea Turtles," Nature 380 (1996): 59–61. Copyright 1996 Macmillan Magazines Ltd. Reprinted with permission.

Other Animals

In the meantime, scientists were suggesting that a host of other creatures might also be using Earth's magnetic properties. Whales, salmon, yellow tuna, honeybees, butterflies, termites, cave salamanders, and mole rats were all nominated at one time or another as possible users of magnetic cues. Like migrating birds and sea turtles, many of these animals tend to travel in murky habitats where visibility is limited. Among the more persuasive evidence was the discovery of thousands of spiny lobsters migrating together, marching hesitantly in single file 30 miles across the ocean floor. Lobsters trained to feed at the northern entrance of a tank tended to switch locations when magnetic North was shifted in their tank.

FIGURE 6.4. A BABY SEA TURTLE WEARING ITS LYCRA SWIMSUIT.
Reprinted with permission from K. J. Lohmann and C.M.F. Lohmann.

Records of whale strandings have revealed that live whales tend to beach where they experience unusual magnetic fields. Earth's magnetic field is nominally 0.5 gauss (50,000 nanotesla) at 40 degrees north latitude. The field is strongest at the magnetic poles and weakest at the magnetic equator. But fields sometimes vary around particular geological formations. For example, the Kursk magnetic anomaly southwest of Moscow is three times more magnetically intense than the average.

Data collected from the British Isles, Newfoundland, and the east coast of North America indicate that live whales tend to strand where valleys of relatively low magnetic intensity cross coastlines or are blocked by islands. In addition, live whales usually strand outside their normal habitat within one or two days after a magnetic storm, when magnetic fields can change unpredictably by about 2 percent in association with sunspot and solar flare activity.

Chinook salmon, *Oncorhynchus tschawytscha,* and sockeye salmon, *Oncorhynchus nerka,* born in ponds and streams swim while still young down rivers to estuaries and the Pacific Ocean. As mature fish, they migrate back from the ocean to the streams where they were born. Fishery agencies tag and release 40 million cultured salmon yearly with metal labels identifying the place and date of their birth. When sport and commercial fishermen catch the fish, they return the tags to research stations. According to the data collected, 90 percent of the salmon that make it back to freshwater return to their home stream, a remarkable navigational score by any reckoning. Because the Pacific sky is so often overcast—the Seattle area, for example, has 228 cloudy and 83 partly cloudy days annually—scientists have concluded that salmon must be navigating by magnetic cues.

Scientists devised an experiment to test the salmon fry's ability to navigate by means of magnetic cues. Salmon fry swimming north into Seattle's Lake Washington were caught and placed in an enclosed wooden tank. A magnetic coil around the tank changed the direction of magnetic north so that it no longer corresponded with geographic north. At that point, the salmon fry dutifully switched direction and followed the magnet's field. At the same time, their behavior was compared with a group of salmon fry who were swimming in the opposite direction, heading south into Chilko Lake in British Columbia. They too were placed in a wooden tank and surrounded by an artificial magnetic field that rotated magnetic north away from geographic north. Immediately, they too made an abrupt turn so that they could continue heading magnetically south.

In their hives, honeybees communicate with their co-workers by dancing along the eight cardinal points of the magnetic compass. Honeybees tend to orient their combs, their waggle dances, their resting positions, and their hives with magnetic fields. Their behavior depends on the intensity of the field. By performing a waggle dance on vertically hanging sheets of wax honeycomb, a bee tells its fellow workers the angle between the sun and the food source that the bee has just visited. If the hive is laid on its side so that the honeycombs are oriented horizontally, the waggle dancers become confused. Within two or three weeks, however, they are once again dancing along the points of the magnetic compass. When bees begin a new hive, they tend to build the new sheets of comb in the same magnetic orientation as in their parent hive. Fascinated by the effect of magnetic fields on the bees' waggle dance, researchers have tried to locate a magnetic detector in bees and to train the bees to discriminate among magnetic fields.

Over the years, these magnetic responses intrigued many natural scientists. However, most of the information accumulated about animal orientation and magnetic compasses concerned animal behavior. The only convincing evidence, as far as physical scientists were concerned, was the magnets discovered in bacteria and the tightly controlled orientation experiments conducted with birds and turtles. Surprisingly little was known about the iron materials in the animals which could serve as magnetic detectors.

The Materials—Living Synthesis of Iron Minerals and Magnets

It had long been recognized that animals can synthesize mineral compounds. Some animals biosynthesize iron minerals and store them passively in various parts of their bodies out of harm's way. Aquatic mammals called dugongs store excess iron in their livers and coat them with the gold-colored iron compound goethite. Certain sponges also form iron compounds. Some animals that biosynthesize iron compounds have actually found a use for them. Limpets (gastropods) harden their teeth with goethite. Rats store large supplies of iron near their teeth in ferritin iron-storage proteins; then they coat their incisors with the iron, the better to gnaw with. In addition to geothite, animals can biosynthesize several $FeO(OH)$ iron compounds, including lepidocrocite and ferrihydrite.

Most important, scientists discovered that bacteria are not the only creatures that can make magnets. Actual *magnetic* iron particles were discovered in the world's largest chiton, *Cryptochiton stelleri,* a 20-inch mollusk that grows in intertidal zones of the northwest Pacific. The chiton scrapes and files algae off rocks with saber-like teeth arranged in rows on a conveyor belt tongue called a radula. The central teeth on each row are extremely tough because they contain large amounts of magnetite, the hardest known substance that can be biologically synthesized. Like rats, the chitons store large quantities of iron near their teeth in ferritin iron-storage molecules. When Heinz Lowenstam of Caltech discovered this material in 1962, scientists were amazed that any living creature could make magnetic iron at ambient temperatures and atmospheric pressures. The fact that chitons and limpets have evolved two different strategies for handling high levels of iron may suggest medical treatments for helping human patients with too much iron in their bodies. The first magnetite in a mammal was discovered in the head of a common Pacific dolphin, *Delphinus delphis*, in 1981, but this material is not permanently magnetic.

However, in 1988, researchers did discover packets of magnetic particles, linear chains of magnetosomes like those of magnetic bacteria, inside the skulls of sockeye salmon, *Oncorhynchus nerka*. The amount of this magnetite, found even in young sockeye salmon is enough to allow them to detect small changes in the intensity of Earth's magnetic field. The fact that the magnetic particles appear to be crystalline suggests that the magnetite is not haphazardly produced but must be under some sort of precise biological control.

Although biologists had come to accept the fact that a number of species sense magnetic fields, they did not know what sense organ was involved. Is magnetite the basis for this entirely new sense? Humans use eyes to see, taste buds to taste, and ears to hear. If other creatures have a sixth sense, what is its sensory organ? No magnetic-field receptor had ever been identified in an animal. Not knowing how animals sense magnetic fields was akin to knowing that people see without knowing that they have eyes.

Locating the sense organ proved extremely difficult. Magnetic fields are relatively simple phenomena. They vary only in direction and intensity, and they change slowly in both space and time. Thus, to sense them, an animal needs only a few sensory nerves and a rather simple central processing center. Since magnetic fields pass straight through tissue, receptor cells could be spread throughout the body. Single-domain magnetite particles like those found in magnetic bacteria and salmon are extremely small and few and thus difficult to detect in larger organisms. They cannot be detected by conventional light and transmission electron microscopes. The fact that some animals respond to magnetic cues only when in motion further complicates laboratory work.

Particles of single-domain magnetite found in homing pigeons and white-crowned sparrows, *Zonotrichia leucophrys*, hinted enticingly at the identity of the birds' magnetic sense organ. The magnetite particles were located in the birds' neck muscles. Physicist Ellen Yorke of the University of Maryland in Baltimore County calculated that the birds contained enough magnetite to sense the magnetic pull of Earth.

But how could the magnetite orient a bird to fly in a particular direction? The magnets in pigeons and white-crowned sparrows lie close to small organs called muscle spindles. These spindles are encapsulated in specialized muscle cells containing receptors that are acutely sensitive to mechanical stretch. Biologists speculated that Earth's magnetic field could create a force on the magnetite particles and make them move ever so slightly to mechanically stretch the spindle. Then stretch receptor cells would respond to the speed and strength of the stretch and send a biochemical signal to a neuron in the brain.

Were these particles and receptor cells part of a sensory system? Could information be relayed from the magnetic particles to the brain? And, even more difficult, could this sensory pathway be linked to the way an animal responds to magnetic cues? Neurologists could not yet explain how the magnetite particles could act in combination with the stretch receptors to create this neurological system.

In 1998, Michael Walker and his colleagues at the University of Aukland in New Zealand found cells containing magnetic iron in the nose of a rainbow trout, *Oncorhynchus mykiss*. Even more important, they located a nerve that connects the nose with the brain and that responds to changes in magnetic field intensity. To show that trout reacted to magnetic cues, Walker and his team trained four trout to swim to a target and strike it for food. When a strong magnetic field surrounded the target, the trout struck the target repeatedly and were rewarded with food. Conversely, when there was no strong magnetic field around the target, the fish were not rewarded with food and so struck the target far less frequently.

Walker and his team had found a likely mechanism for the animals' magnetic sense. Using magnetite in their noses, the animals can detect magnetic fields, and magnetic field information can be carried to the brain by the nearby neurons. The Wiltschkos were right after all. Some animals do indeed have a sixth sense that we humans are completely unaware of. Thanks to the iron at the core of our planet, some animals "smell" North.

The ability of migrating birds, sea turtles, whales, salmon, bees, and other creatures to actually sense and orient by iron-generated magnetic fields is a spectacular example of the biological diversity that enriches our lives.

But iron is also a vital player in global mechanisms that nurture the plants, soils, and oceans blanketing the surface of our planet. Iron-based nitrogen-fixation processes fertilize the land while iron nourishes the seas. Thus, species that use magnetic iron to traverse the globe are, in turn, nurtured by vast iron-based systems that feed the oceans and make nitrogen available for amino acid building. So we turn now to the plant world and the powerful role that iron plays in the very seas and soils that nourish us and all living creatures on planet Earth.

7

Iron and the Planet's Ecosystem
SEAS AND SOILS

Sprinkling iron like salt and pepper over a barren region of the South Pacific, oceanographers made the ocean's surface explode with plant life. From their research ship, scientists stared as clear blue water turned as murky green as a duck pond. Fish and marine birds, turtles, and bacteria flocked to feast on the bonanza. In nine days, the iron made the equivalent of fifty trees grow 120 feet tall—until the plants just as abruptly died and sank. Never before had iron's immense power over plant growth been as dramatically illustrated. This controversial experiment was conducted in 1995 to test the idea that fertilizing the oceans with iron could increase marine plant growth and thus remove carbon dioxide (CO_2) from the atmosphere to alleviate the "greenhouse effect," which contributes to global warming. Although it turns out that seeding the world's oceans with iron will apparently not accomplish this goal, a deeper understanding of iron biochemistry may someday help feed the planet's hungry.

Two of the most fundamental life processes—photosynthesis and nitrogen fixation—depend on iron. In photosynthesis, iron directs the assimilation of the sun's energy. In nitrogen fixation, iron gives leguminous plants, via symbiosis with soil bacteria, the ability to turn the nitrogen in air into amino acids for protein biosynthesis. Together, nitrogen fixation and photosynthesis paint a powerful picture of iron's ability to nourish our planet.

Iron Seeding

The original purpose of the 1995 iron-seeding cruise was narrow in its scope. Its goal was to answer a riddle, a long-standing biological puzzle nicknamed "HNLC," for high nutrient–low chlorophyll areas

THE GREENHOUSE EFFECT

Earth is like a giant greenhouse, with an atmosphere for its walls and roof. Electromagnetic radiation of all wavelengths from the sun enters Earth's atmosphere. The energy of the radiation is most intense at a wavelength of 4.83×10^{-7} meters, which is the green portion of the visible spectrum. Most of this radiation passes easily through the atmosphere to warm Earth.

Earth, in turn, re-radiates energy at longer wavelengths. The most intense re-radiated energy consists of invisible, infrared rays (so called because their wavelengths are longer than the red of the visible spectrum). Some of this infrared "heat energy" is absorbed by the carbon dioxide in the atmosphere and gets trapped there. Any increase in the amount of carbon dioxide exacerbates the effect and contributes to global warming. The burning of fossil fuels has increased the amount of carbon dioxide in the atmosphere.

Only plants and a few carbonate-producing organisms such as corals can safely return atmospheric carbon to storage by converting it to carbohydrates and oxygen. The living world stores carbon as wood and leaves on land and in soils; as limestone in oceans; as organic carbon, coal, and petroleum in marine sediments; and as dissolved organic carbon in water. Thus, clean nuclear power, solar energy, and biological forces are our only allies in fighting the greenhouse effect produced by our unfettered use of fossil fuels.

of the ocean. Twenty percent of the ocean surfaces—the Gulf of Alaska, the ice-free Southern Ocean around Antarctica, and the eastern equatorial Pacific—abound in light and in major nutrients, including nitrates, phosphates, and silicates, but contain little chlorophyll. More recently, David Hutchins of the University of Delaware has discovered HNLC regions close to shore, for example, off Big Sur, California. In spite of the presence of light and nutrients, all these waters are remarkably infertile, a clear and brilliant blue instead of chlorophyll green. Why are they not filled with phytoplankton, the microscopic floating plants that form the bottom of the marine food chain?

Oceanographers were especially intrigued by HNLC regions because these regions may have played major roles in past climate changes.

One Scientist's Chutzpah

When the late John Martin, an expert in measuring trace metals in ocean water, heard about unused nutrients off Antarctica in 1986, he wondered if the phytoplankton there could be suffering from iron "anemia." Martin was a scientist at Moss Landing Marine Laboratories, a consortium of seven University of California oceanographic programs. Almost all the plants on Earth depend on iron to direct photosynthesis, which incorporates carbon dioxide (CO_2), sunlight, and water into sugars for energy and oxygen. In photosynthesis, sunlight gives electrons in the plant energy to jump about and cause chemical reactions; the electrons use iron atoms as parking spaces during this process. Plants, like humans, store iron atoms in ferritin molecules, each one of which can hold as many as two thousand iron atoms. Inside plant cells, ferritin molecules are particularly plentiful in chloroplasts, the minute chlorophyll-containing structures where photosynthesis occurs. Without iron, leaves cannot photosynthesize, and their lush green color turns sickly yellow.

As Martin pointed out, if phytoplankton were deprived of iron, they would not be able to use the nutrients in seawater. Martin's suggestion that a lack of iron might cause the infertility problems of oceans was met with "great skepticism," the polite term that scientists often use for utter disbelief. Martin's trademark was his dogged determination, though, and he began collecting data to support his theory. Seawater contains so few free iron atoms that measuring them proved daunting. Iron ships are horrendously polluted laboratories for anyone who wants to measure extremely small amounts of iron. But Martin managed to analyze his water samples in clean rooms on board ship, using Teflon-coated sampling gear, ultraclean laboratory equipment, and the like. He was reviving a 1930s suggestion that ocean surfaces must be short of iron wherever they fail to get resupplied with iron from land masses.

First, Martin realized that the iron in surface seawater must come from land, either from runoff or from dust blown in from arid soil.

Next, he and his co-workers discovered offshore areas with extremely low amounts of iron, a tenth of what marine plants need to grow. Then he observed that the amount of atmospheric dust in the barren areas of the Antarctic and equatorial Pacific is lower than anywhere else on Earth. In contrast, the North Atlantic is continually enriched by iron dust blown in from the Sahara desert.

If Martin had stopped there, the role of iron in infertile seawater would have remained an intriguing scientific puzzle: no more, no less. But Martin was charismatic and loved a good argument. Wheelchair bound from the aftereffects of childhood polio, he had a far-ranging imagination and the chutzpah to match. Martin quickly politicized the issue.

Attending a lecture at Woods Hole Oceanographic Institution in 1988, Martin put on his best Dr. Strangelove accent and more or less facetiously suggested that "with half a shipload of iron he could create an ice age." Adding 300,000 tons of iron to the Southern Ocean should make it bloom with phytoplankton. When the plants died and sank to the seafloor, Martin estimated they would take with them 2 billion tons of carbon dioxide, one of the prime culprits in global warming. Sink enough carbon dioxide and the world should cool off, he said, noting in support of this theory that atmospheric dust during the last ice age had been fifty times richer in iron than current levels are.

Sitting outside on the lawn sipping beer after Martin's remark, colleagues were skeptical about iron's role in the low phytoplankton population; it seemed more likely that other factors were keeping phytoplankton populations low. Floating microscopic animals called zooplankton might be eating the plants. Cold temperatures or a lack of light might affect their growth too. So, relaxing in the July sunshine, the crowd of oceanographers decided that Martin was joking; no one present thought he was serious about fertilizing the ocean.

But Martin was serious. He began lining up scientific support for an ocean-going experiment to test iron's ability to make phytoplankton grow and use up enough carbon dioxide to reverse global warming. Within three years, he had secured approval for a small-scale experiment from a National Academy of Sciences committee and from the American Society of Limnologists and Oceanographers, the main professional organization of scientists in the field. In one of his scientific articles in *Nature,* he declared boldly: "Oceanic iron fertilizing aimed at the enhancement of phytoplankton production may turn out to be the most feasible method of stimulating the active removal of greenhouse gas CO_2 from the atmosphere, if the need arises." Normally, such speculative, unproved comments do not appear in a scientific article, and their publication in one of science's most prestigious journals helped give Martin's proposal the imprimatur of respectability.

Tampering with the ocean is anathema to many, if not most, oceanographers, however. Oceanography is a new science, and oceanographers have not even compiled all the ocean's vital statistics, much less explained the complex relationships of the ocean's ecosystem. Sallie W. Chisholm, a professor in the biology and engineering departments at the Massachusetts Institute of Technology, argued with Martin about the ethics of seeding the seas to reduce global warming. "He was cynical about mankind," she said later. "He felt that obviously this would be a foolish way to go, that humans should get their act together and stop emitting so much carbon dioxide. But he didn't believe that we would get our act together."

On May 20, 1990, a *Washington Post* reporter with his ear to the ground wrote a front-page article that began, "Scientists trying to battle the 'greenhouse effect' have seriously proposed dumping hundreds of thousands of tons of iron into the ocean to create giant blooms of marine algae that could soak up much of the excess carbon dioxide believed to be responsible for global warming. If the massive scheme is carried out, researchers say, it would be among the greatest manipulations of nature ever attempted. The proposal to dump iron into the oceans has been given quiet endorsement by a special panel of the National Research Council, an arm of the National Academy of Sciences and Engineering, which is chartered by Congress to advise the government on scientific and technical issues."

"The cat was out of the bag," as Martin put it. What he called "the Geritol solution"—pumping iron, dosing oceans to cool off the planet, producing an ice age—these were issues that anyone could understand. Articles and letters to the editor filled the popular press and scientific journals. A particularly cogent summary of these issues appeared in the Toles comic strip (see figure 7.1).

FIGURE 7.1. AN EXPLANATION OF IRON SEEDING IN THE OCEANS IN A TOLES CARTOON.

Toles © 1990 The Buffalo News. Reprinted with permission of UNIVERSAL PRESS SYNDICATE. All rights reserved.

Highly politicized by now, the experiment chugged along on two levels. The first ranged over political, environmental, and ethical questions. And the second—a distant second at that—focused on the scientific HNLC puzzle: Was an iron shortage responsible for infertility in the ocean?

A group of oceanographers at the University of Washington issued a statement to major newspapers, scientific organizations, government officials, and committees: "A few seemingly clever scientific manipulations of a natural marine (or terrestrial) ecosystem cannot be expected to solve a complex problem of global proportions that

has been a century in the making. . . . Controlling human exacerba-
tion of the global greenhouse problems depends primarily upon an
enlightened U.S. citizenship, since we continue in apparent apathy to
use more fossil fuel and energy per capita than any other people on
Earth." "Most thought," as Sallie Chisholm said, that there was "very
little evidence that it would work and that even if it did work it was
inappropriate. . . . The arrogance of human beings is just astounding."

The Expeditions

Armed with the official endorsement of the oceanographic community, however, Martin persuaded the National Science Foundation to
fund an experiment in the open ocean. In November 1993, shortly after Martin's death from cancer, the research vessel *Columbus Iselin*
sailed 300 miles (500 kilometers) south from the Galapagos Islands
off Ecuador into an infertile area of the equatorial Pacific. Chemists
Kenneth Johnson and Kenneth Coale, both from Martin's laboratory
at Moss Landing, had planned the experiment.

Stationing a buoy filled with electronic and chemical testing equipment in the center of their test site, they pumped about half a ton
(400 kilograms) of iron sulfate into the wash of the ship's propellers
as the vessel covered a 25-square-mile (64-square-kilometer) patch
around the buoy. Mixed with the iron was a chemical tracer that enabled the scientists to follow the floating iron patch for nine days.

On board the ship were physical oceanographers testing the water's temperature and salinity; biologists looking at plants, chlorophyll, algae, and zooplankton; chemists measuring pH and the levels
of iron, carbon dioxide, and the tracer chemical; and biophysicists
analyzing the phytoplankton. Overhead flew a NASA aircraft with an
airborne optical laboratory.

The results were startling and almost instantaneous. Although
1 million tons of phytoplankton had been produced by the end of the
third day, something had clearly gone wrong. A mass of more buoy-
ant, unfertilized water had slid over the iron-enriched patch, burying
it 90 feet deep. Sampling became difficult. Plant growth ceased.

There was another problem too: the amount of carbon dioxide
consumed was quite small, a mere 10 percent of Martin's estimates.
The iron did indeed stimulate plant growth, but it did not consume

much carbon dioxide from air. As marine biologist Richard Barber, chief scientist on the cruise, noted wryly, "Apparently the phytoplankton in the patch hadn't read the literature." The expedition proved that iron starvation creates barren regions in the oceans, but it did not prove Martin's second hypothesis: that iron seeding would halt global warming.

In June 1995, a bigger research ship, the *R. V. Melville,* sailed into the Pacific loaded with even more equipment and scientists than before. This time the iron was sprinkled over the water in three batches, and again the ocean bloomed with trillions of organisms in "almost biblical proportions," as one scientist reported in awe; another said, "It was like driving through the Mojave Desert and finding a rain forest." The iron produced two thousand times its own weight in plants, clogging and ripping the vessel's nets. The population of diatoms, a type of unicellular algae, multiplied eighty times. Carbon dioxide levels in the seawater plunged 15 percent.

Monitoring the tracer chemical, the research vessel followed the 25-square-mile patch as it drifted 900 miles (1,500 kilometers) across the Pacific. Some of the scientists were delighted, and others burst into tears. Deckhands asked scientists, "Did we do this?" Kenneth Coale admitted, "It was a profoundly disturbing experience for me."

But this bloom was short-lived too. Within a week of the last iron treatment, the patch had returned to normal: barren and blue. As before, the iron eventually sank, and bacteria ate the plants.

Given the rudimentary state of our knowledge about the oceans, it is still not known whether larger-scale efforts at iron fertilization will reduce the effects of global warming. Some computer models have suggested that fertilizing the ocean with iron would have only a small and transient effect on carbon dioxide concentrations in the atmosphere. Dosing with iron is likely to increase the conversion of carbon dioxide to organic carbon by only about 10 percent, far too little to cure the problem. Moreover, when the phytoplankton bloomed, marine bacteria appeared to gobble up the vegetation. Competing directly with the phytoplankton for iron, they consumed the metal twice as fast as the plants did.

In fact, later experiments suggested that an iron cure could be worse than the disease, that it might even exacerbate global warming. The decay of the phytoplankton bloom could use up enormous amounts of oxygen, turning ocean regions into polluted ponds, killing marine life, and generating huge amounts of methane gas, a global warming chemical even more potent than carbon dioxide. Phytoplankton and marine bacteria also compete for carbon. Phytoplankton absorb carbon dioxide and convert it to carbon. Marine bacteria take up this carbon in turn and convert it back to carbon dioxide Scientists agree that until the iron and carbon competitions are quantified, the results of iron-seeding experiments will be quite unpredictable.

Scientists have reacted in two different ways to the iron-seeding experiments. Some have suggested fertilizing the Gulf of Mexico and the area around the Marshall Islands to increase fish production. Others are concerned that too little is known about the ecology of the oceans to tinker with them; these researchers argue that attempts to circumvent the greenhouse effect by seeding the oceans with iron are impractical, premature, and even potentially dangerous to the planet's ecosystem.

Nitrogen Fixation

The greenhouse effect is not the only challenge to Earth's ecosystem. In the near term, the planet's population is expected to almost double by the year 2050, primarily because of present high birthrates in emerging economies. The key to feeding exploding populations in the developing world does involve iron, but not by spreading it around the land. The key is genetically manipulating nitrogen fixation, the extraordinarily iron-intensive process by which plants produce amino acids. Nitrogen fixation is nature's only method for providing the nitrogen atoms that plants need to make amino acids. Nitrogen fixation starts with atmospheric nitrogen and ends up with ammonia, the crucial ingredient for producing amino acids.

Most nitrogen fixation on Earth—60 percent of it—is accomplished by lowly microorganisms and plant roots in soil. Although they too have great difficulty synthesizing ammonia, they manage to do it at ambient temperatures and pressures. No higher organism has evolved the skill to fix nitrogen. Given the chemical difficulties involved, it is not surprising that only a few genera of bacteria have evolved the ability. By far the biggest share of plant-produced organic nitrogen is made by a symbiotic partnership between a few plants and their bacteria. In fact, 40 percent of the nitrogen fixed by bacteria and virtually all the fixed nitrogen used by cultivated plants come from one particular symbiotic relationship: that between *Rhizobium* bacteria and leguminous plants, which include beans, peas, peanuts, soybeans, alfalfa, clover, kudzu, locust, acacia, and vetch (see figure 7.2). Actually, the process is even more limited because nitrogen fixation occurs primarily in the roots of these plants and because each legume species fixes nitrogen with only one particular *Rhizobium* species. Thus, for example, the *Rhizobium* species that allows alfalfa to fix more than 220 pounds (100 kilograms) of nitrogen per acre per year cannot help beans or peanuts fix nitrogen at all. Compared with nature's production of organic nitrogen in soils, all the fertilizer factories in the world fix only 25 percent of the nitrogen available for plant use. If bioengineering could copy into cereal crops the ability to fix nitrogen, they could be grown without expensive nitrogenous fertilizers.

What is iron's role in this process? While photosynthesis takes place in plant leaves, nitrogen fixation occurs primarily in the roots of legumes. Two ancient classes of iron proteins—iron-sulfur proteins and, surprisingly, a plant hemoglobin—enable the *Rhizobium* bacteria in legumes to break down nitrogen molecules into nitrogen atoms for use in making essential amino acids and proteins. A monstrously large iron-sulfur protein, the enzyme nitrogenase, is composed of two smaller iron proteins that together contain as many as thirty iron atoms. All nitrogen-fixing organisms, including *Rhizobia* bacteria, use nitrogenase; and some use not one but several varieties of nitrogenase, including iron-molybdenum complexes and iron-vanadium complexes. The plant hemoglobin, similar to the enormous molecule that is used to transport oxygen through the blood of vertebrates, is called leghemoglobin. It is produced by the symbiotic relationship between legumes and bacteria. In addition to nitrogenase and leghemoglobin, legumes and their companion *Rhizobia* bacteria also employ many other smaller iron-sulfur proteins.

The first step for any legume that needs to fix nitrogen is to acquire iron from the soil. Armed with special chemicals that are not well understood, the roots of most plants can reduce the ferric iron in soil to the ferrous iron that plants can use. When grown in iron-deficient soils, some plants—notably grasses such as barley, wheat, oats, rye, and rice—make large numbers of iron-grabbing siderophores. These

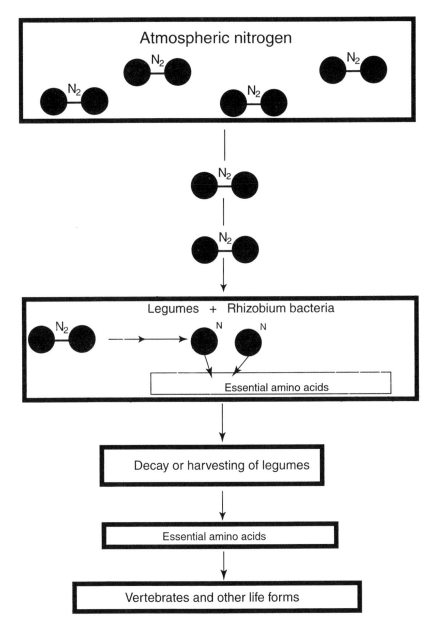

FIGURE 7.2. AN OVERVIEW OF NITROGEN FIXATION.
Atmospheric nitrogen molecules (N_2) are converted to nitrogen atoms (N) by means of the symbiotic relationship between a legume and its Rhizobium *bacterium. The legume then uses the single nitrogens to make essential amino acids. When the legumes are harvested or decay, these essential amino acids enter the food chain.*

molecules, which are like those described in chapter 3, are called phytosiderophores because they are made by plants, not bacteria. Rice, for example, is extremely susceptible to iron deficiency and produces a phytosiderophore known as mugineic acid. Plant sidero-phores are much smaller than those made by bacteria, but their active, iron-grabbing centers are quite similar. Plant siderophores are particu-larly skilled at extracting iron from alkaline soils, which strongly bind iron compounds. *Rhizobium* species not only make their own sidero-phores but also use siderophores produced by other soil bacteria.

Courtship

Once supplied with iron, the legumes and their bacteria begin a stately sort of conversational courtship, signally each other chemically and collecting information that will allow their unique marriage to pro-ceed. The conversationalists in this affair are an odd couple, to say the least: the host (the plant) is a member of the eucaryotic kingdom, and the invited guest (the bacterium) is a procaryote. It is a bit like a frog phoning a mushroom or a seal romancing a fern.

This elaborate chemical courtship allows the bacteria to modify what is basically a disease process in order to enter the legume in a mutually beneficial manner. The plant needs to be infected by a *Rhizo-bium* in order to fix inert nitrogen. But only a specific *Rhizobium* spe-cies will do; another might turn out to be pathogenic or parasitic. Thus, the plant and bacterium must recognize and signal each other before the action can begin. Once the process is underway, they con-tinue to communicate back and forth, each initiating a reaction from its partner.

The legume begins by chemically signaling its *Rhizobium* bacteria to enter its root hairs. Reaching across kingdoms, the plant's signal turns on a nodule-making gene in the bacteria. In response, the bac-teria's *nod* gene disables the plant's defenses so the bacteria can en-ter the plant (see figure 7.3). Back and forth they go, a member of one kingdom forcing a change in the other's kingdom. Specifically, the bacterium's nodulating gene makes the bacterium produce enzymes that deform the plant's root hairs. Curling around the bacterium, the hairs spin a long and sinuous "infection thread." Traveling up the thread, the rod-shaped *Rhizobium* bacterium enters the plant's cells.

Legume

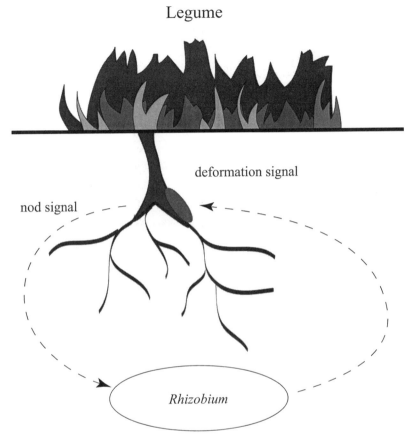

FIGURE 7.3. THE COURTSHIP ACROSS KINGDOMS.

The legume signals its Rhizobium *bacteria; the bacteria respond with a deformation signal, and nodulation of the plant root begins.*

Dividing rapidly, the plant's cells eventually form nodules that look like swellings. These nodules are safe houses where nitrogen fixation can occur. Inside the nodules, the bacteria make special compartments called bacteroids. Inside these bacteroids, the proteins needed for fixation are synthesized (see figure 7.4). The nitrogen from the roots is then assimilated into amino acids, which are exported to the rest of the plant for making proteins.

Five or more iron compounds, including nitrogenases and leghemoglobin, occupy the heart of the fixation process inside the nodule.

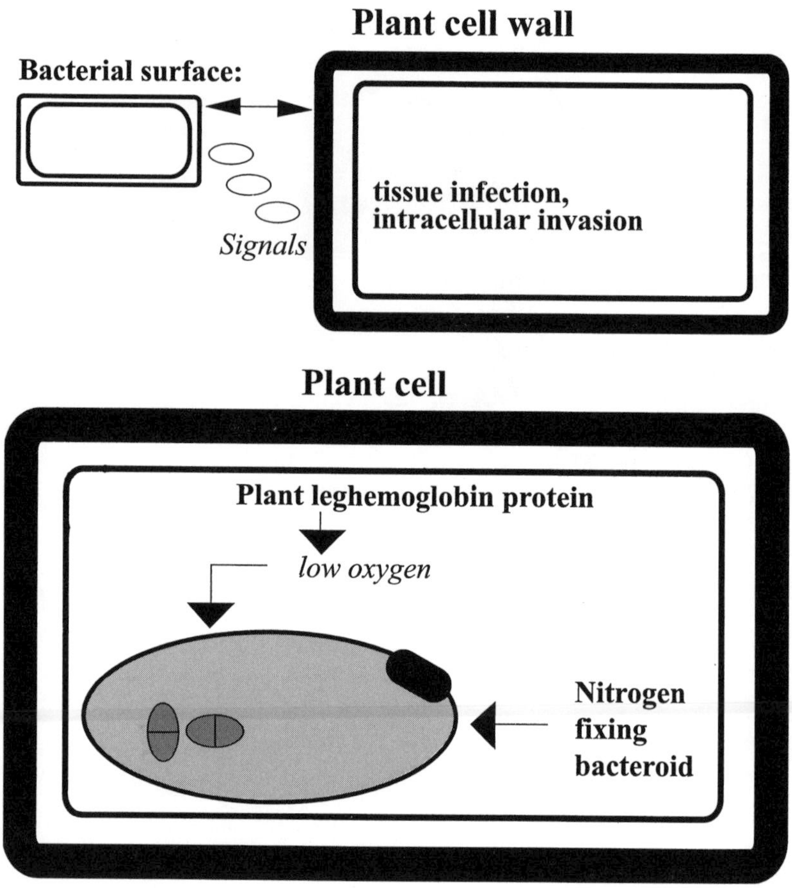

FIGURE 7.4. THE CHEMISTRY OF NITROGEN FIXATION.
A legume and its bacterium signal back and forth to turn atmospheric nitrogen into ammonia. First, the legume sends a nod *signal to its* Rhizobium *bacterium. The bacterium responds by sending a deformation signal to the root of the legume. Inside the nodule (bottom part of diagram), the bacterium creates a special compartment where the chemistry of fixation occurs. This chemical process requires low oxygen levels, which are controlled by leghemoglobin. A large iron-containing enzyme called nitrogenase facilitates the process. Adapted from S. R. Long and B. J. Staskawicz, "Prokaryotic Plant Parasites,"* Cell *73 (1993): 921–935. Copyright © Cell Press. Reprinted with permission.*

Some nitrogen fixation systems are catalyzed by not one but several varieties of nitrogenase.

Leghemoglobin, the only hemoglobin known in plants, has been found in the nodules of every nitrogen-fixing legume. Leghemoglobin is closely related to the hemoglobin in vertebrates. Cut a legume nodule open and put it under a light microscope, and the red color of hemoglobin is clearly visible. Structurally, however, leghemoglobin is actually more closely related to vertebrate myoglobin than to hemoglobin. The plant version consists of an iron-centered heme and two amino acid chains, instead of hemoglobin's four. Each legume produces its own unique sequence of amino acids in the globin chains, and thus each type of leghemoglobin has a slightly different affinity for oxygen. The details of how leghemoglobin is made are not known. In the culmination of their odd relationship, the bacterium apparently produces the heme while the plant produces the globins.

However it is made, leghemoglobin fills the nodules, constituting 40 percent of all the soluble protein there. The concentration of leghemoglobin is actually higher in some legume nodules than the concentration of myoglobin in vertebrate muscles. The leghemoglobin molecules are so numerous in the nodule that they bind 99 percent of the oxygen there. This is fortunate, because nitrogenase can fix nitrogen only in the presence of very small amounts of oxygen.

Trapping most of the oxygen molecules before they can reach the *Rhizobium,* leghemoglobin allows the bacteria to have their cake and eat it too. Thanks to the leghemoglobin, the bacteria can have an efficient, oxygen-based metabolism and keep the nitrogen-fixation process in the bacteroid almost oxygen free. Leghemoglobin is surely one of the primary benefits that the bacteria get out of their symbiotic marriage with legumes.

Similarities between plant and animal hemoglobins suggest that they have a common ancestor, probably from an archaea that developed first more than 3.5 billion years ago. Perhaps genes from the common ancestor of all hemoglobins will be found in contemporary brine-dwelling archaea, the only surviving archaea known to have aerobic respiration.

The marriage between legumes and *Rhizobium* bacteria is the most efficient natural nitrogen-fixing system, but it is not the only one. In the rice paddies of China and Vietnam, rice is intercropped with a small aquatic fern that works with the aquatic cyanobacteria *Anabena*. To fix nitrogen in symbiosis with lichens, liverworts, and ferns, albeit at a tenth the rate of legumes, the cyanobacteria provides nitrogen for the ferns, which decompose in the soil and provide nitrogen for the rice.

Using some chemical sleight-of-hand, primitive cyanobacteria have even developed a way to make oxygen at the same time that they fix nitrogen anaerobically. In photosynthesis, cyanobacteria use water and the energy from sunlight to convert carbon dioxide to organic compounds and to liberate oxygen. But they can also use energy from sunlight to fix nitrogen. They keep the oxygen from poisoning the nitrogen fixation process by separating the two processes in either space or time. Spatially, the bacteria build heterocysts, structures with thick walls that keep oxygen out while nitrogen fixation is occurring inside. Other cyanobacteria species use a primitive version of time-sharing by fixing oxygen during the day and fixing nitrogen at night.

About 15 species of soil microorganisms fix nitrogen on their own, without forming any symbiotic relationship with a plant. Although free-living bacteria fix about 100 times less organic nitrogen than *Rhizobium* bacteria fix with legumes, free-living bacteria have taught biologists a great deal about fixation chemistry.

The Scarcity of Nitrogen Atoms

Single nitrogen atoms are crucial for the production of amino acids, each one of which contains at least one nitrogen atom. Although there are only twenty common amino acids, they combine in various ways to assemble at least sixty thousand different proteins, the long-chain molecules that carry on metabolism in living cells. While vertebrates can synthesize eleven of the twenty amino acids from food, only nonvertebrates can manufacture the other nine. These nine are called the essential amino acids. Humans cannot make any of these nine amino acids, but plants can, although even they must use the complex and energy-intensive technique of nitrogen fixation.

Unfortunately for plants, nitrogen has the odd property of being both abundant and scarce at the same time. Molecules of nitrogen (N_2) account for 78 percent of Earth's atmosphere, but they are so chemically inert that they are useless as far as the living world is concerned. The nitrogen in our air occurs in the form of molecules made of two nitrogen atoms. Amino acids, on the other hand, are made only from single atoms of nitrogen that are bound to hydrogen or carbon. Thus, before a molecule of atmospheric nitrogen can be used by a living organism, it must be split into two nitrogen atoms. By itself, however, a nitrogen atom is highly reactive and wants to be attached to another atom. In its search for bonding partners, it can tear molecules apart. But safely bound to other elements in compounds, nitrogen atoms are both available to plants and able to participate in far milder reactions.

A shortage of nitrogen compounds, especially ammonia (NH_3) or nitrate (NO_3), is one of the most common limitations on the growth of plants and animals. The nitrogen used by living organisms is generally called organic nitrogen.

The amount of organic nitrogen in our biosphere is constantly being depleted by bacteria that use nitrogen as an energy source. These bacteria turn organic nitrogen back into nitrogen molecules (N_2). Harvesting plant crops to feed people also depletes soils of organic nitrogen. Thus, organic nitrogen must be constantly replenished, or world food production will decline. To feed the world's rapidly expanding population, more organic nitrogen will have to be incorporated into the food chain. Realistically, many biologists say, this can be accomplished only by helping plants fix more nitrogen.

Fixing nitrogen, however, is a remarkably difficult chemical process. As a result, the number of organic nitrogen atoms in the world is rather stable. In fact, many of them have been recycling through the food chain for eons. Animals absorb nitrogen compounds in their food and excrete it as ammonia, urea, and uric acid, thus fertilizing more plants. Feces have been used for centuries as fertilizer.

Both fertilizer factories and *Rhizobia* bacteria in nitrogen-fixing plants fix nitrogen by breaking apart the two atoms of molecular nitrogen into single atoms and joining them to hydrogen atoms to make ammonia (NH_3). First, the bond between the nitrogen atoms in the dumbbell-shaped N_2 molecule must be broken. Then three hydrogen atoms must be bound to each nitrogen atom. Finally, because ammonia is toxic to most plants, it is usually combined with another element to make a salt such as ammonium nitrate (NH_4NO_3).

Formidable amounts of energy are needed to fix nitrogen atoms into molecules that living organisms can use. Energy-intensive processes such as lightning and the ultraviolet radiation in sunlight produce only about 15 percent of single nitrogen atoms. Even a plant must use between twelve and twenty-four molecules of ATP to convert one molecule of N_2 into two molecules of ammonia. ATP is the carrier of chemical energy in living organisms. Industry produces 25 percent of the world's supply of organic nitrogen, mainly in the form of fertilizers and explosives. About 74 billion tons of atomic nitrogen are made commercially by the Bosch–Haber process, which employs an iron catalyst, high temperatures, and high pressures. Because natural gas and petroleum supply the energy used in the process, the price of artificial fertilizer varies with the price of fossil fuels.

The cost of artificial fertilizer is one of the main factors limiting the food supply of poor emerging and Third World nations. If scientists could genetically copy the nitrogen-fixing ability of legumes and *Rhizobia* into cereal crops, grains could also produce the nine essential amino acids. This is unlikely to occur without a deep understanding of the iron biochemistry involved in fixation.

Mere calories will not be enough to nourish the world's exploding population, however. While oceanographers are just beginning to study iron's potency in barren areas of the oceans and while our understanding of iron's power to restore nitrogen-depleted soils is incomplete, evidence is accumulating that iron deficiency in childhood can cause lifelong neurological deficits. The need for iron to save the minds and bodies of young people around the world is all too well documented.

8

Feeding the World's Poor
IRON DEFICIENCY

Human beings are so dependent on iron for their growth and development that iron-poor diets cause one of the world's most prevalent nutritional problems. Iron deficiency, the most common consequence of malnutrition, affects an estimated two-thirds of all children and women of childbearing age. The incidence of iron deficiency peaks during infancy, and an estimated 20 to 25 percent of the world's babies have iron deficiency anemia, the most severe form of iron deficiency. But infants are not the only group affected. In addition, 42 percent of the world's women and 26 percent of its men are anemic, according the World Health Organization (WHO).

Tragically, iron deficiency anemia can irreversibly damage a child's long-term intellectual development. Studies of anemic infants and schoolchildren around the world have consistently shown that iron deficiency can irreparably impair their learning and motor skills. Iron supplementation can correct their anemia but cannot fully restore their cognitive function. The majority of studies also report adverse consequences from mild to moderate iron deficiency and anemia. In addition, children with iron deficiency are more susceptible to lead poisoning, which causes its own set of neurological disorders.

Dietary iron shortages affect many of the world's poor, whether they are in emerging nations or in the United States, Japan, and Europe. In the United States, between 9 and 11 percent of toddlers, adolescent girls, and women of childbearing age are iron deficient, according to a 1997 federal survey. Approximately one-third of them have such low iron stores and hemoglobin levels that they become anemic. The incidence of anemia has declined markedly in the United States since the 1960s, but the condition is still relatively common among poor black and Hispanic women who have four or more children and a high school education or less. At the same time that many poor women and children are iron deficient in the United States, approximately 110,000 American youngsters have been poisoned since 1986 because they confused iron pills with candy. Of these, 35 have died. Federal law now requires childproof packaging for the high-potency iron supplements aimed at pregnant women.

Iron deficiency is most severe during childhood when growth is rapid, during pregnancy, and during a young woman's childbearing years, because of menstruation. Iron-deficient women experience higher than normal rates of maternal mortality in childbirth, have more miscarriages, and bear more premature babies. In India, the maternal death rate among severely anemic women is 20 percent; in Nigeria, it is up to 50 percent. The babies of severely iron-deficient mothers begin life with lower iron stores and experience higher rates of death, illness, and learning disorders. Iron-deficient youngsters get more bronchitis, gastroenteritis, diarrhea, and respiratory infections, and their meningitis is more apt to be fatal.

Only if women begin their pregnancies with adequate stores of iron can they and their babies be assured of adequate iron levels after birth, concluded "Prevention of Iron Deficiency," a 1998 report written by Fernando E. Viteri of the University of California at Berkeley for the National Academy of Sciences's Institute of Medicine. The report concluded that in developing nations "Iron deficiency is the rule" among both teenage mothers and women with frequent pregnancies because so much iron is lost during pregnancy. Teenagers are especially susceptible because they have not had time to build up their iron stores depleted by their adolescent growth spurt.

Unfortunately, detecting and treating iron deficiency are difficult. Its effects remain largely hidden until iron levels drop very low. Adding to the problem is the fact that women in developing nations are often "grossly neglected," according to the National Academy of Sciences (NAS). For example, the "lack of political will and program support in Latin America are clearly evident," the study noted. Medically, there are problems too. Treating a severely anemic pregnant woman and providing her fetus with enough iron at the same time are extremely difficult to accomplish in nine months. Thus, WHO, the United Nations University, and UNICEF (the United Nations International Children's Emergency Fund) have concluded that iron supplementation programs in developing nations should begin when girls are adolescents, long before they get pregnant. They, as well as women of childbearing age, should take 60 milligrams of iron weekly for several months before pregnancy. Such doses cause fewer side effects than larger doses, and excess iron is not absorbed by the body. While nutritionists advise Westerners to take iron supplements regularly only under the supervision of a health-care professional, the NAS recommends that emerging nations distribute iron supplements through community-based organizations instead of medical clinics, because the cost would be lower.

Breast milk is internationally accepted by physicians and nutritionists as the best food for infants. Infants fed breast milk for four to six months have healthier iron stores than formula-fed babies. Although both cow's milk and breast milk have only trace amounts of iron, the iron in cow's milk is not bio-available, whereas 49 percent of the iron in breast milk is available to the nursing infant. In addition, some hypothesize that lactoferrin, a transferrin-like molecule that binds two atoms of iron in breast milk, may help halt the spread of disease-causing organisms in the milk. In 1997, the American Academy of Pediatrics recommended that infants be breast fed during their first year of life. The only acceptable alternative to breast milk, according to the academy, is iron-fortified infant formula. After six months, children need more iron than either breast milk or unfortified formula can provide, and iron-rich solid foods should be introduced gradually at that age.

People with inadequate iron stores cannot produce enough red blood cells or enough hemoglobin in those cells. An anemic person has fewer red blood cells than normal, and the cells are smaller and paler. Anemia causes fatigue, palpitations, dizziness, headache, rapid heart rate, shortness of breath, and an inability to keep warm in cold temperatures. According to the NAS, iron deficiency can cause "impaired physical growth, compromised cognitive development, short attention span and impaired learning capacity, reduced muscle function and energy utilization, decreased physical activity and lower work productivity, lowered immunity, increased infectious disease risk, and impaired fat absorption (most probably including fat-soluble vitamin A)." Iron deficiency also impairs thyroid function.

Historically, anemic people were often described as backward and apathetic. The Confederacy's defeat during the American Civil War has been blamed in part on the listlessness of its poor white soldiers, many of them anemic because of hookworm parasites. As late as 1900, many teenage factory girls in the United States suffered from chlorosis, such an extreme form of anemia that their skin appeared green. Recent studies of Guatemalan laborers on sugar and coffee plantations and of Indonesian road construction workers and rubber tappers showed that they worked more productively after taking iron supplements.

The average person loses approximately 1 milligram of iron each day in shed skin, urine, and bile. To remain healthy, adults need to maintain 3 to 4 grams of iron in their bodies at all times. This is enough to provide most adult males and postmenopausal women with roughly 500 milligrams of iron per liter of blood. However, menstruating and pregnant women and people living at high altitudes need somewhat more iron. Women lose on average about 20 milligrams of iron during each menstrual period, and 10 percent of women lose triple that. Women lose about 740 milligrams of iron during pregnancy and during and after delivery. Intrauterine contraceptive devices, which cause uterine bleeding, also result in iron loss.

Hookworms and other parasites contribute to iron deficiency in more than 1 billion men and women in developing areas of the world. Hookworm itself, largely eradicated from the United States and industrialized nations, causes gastrointestinal bleeding. Although they are rarely fatal, even mild hookworm infections can devastate women of childbearing age, especially pregnant women, according to the NAS. In tropical and semitropical climates, hookworm larvae enter the body through the soles of bare feet and end up in the small intestine, where they eat blood. In the long term, hookworm is eradicated by the adequate removal of fecal wastes and the wearing of closed shoes. In the short term, effective, inexpensive, and safe deworming medications are now available.

To replace lost iron, people must eat food that has bio-available iron in it. Overall, the recommended daily dietary allowance for iron is about 10 milligrams for adult men and 15 milligrams for menstruating women. Only about 10 percent of that is absorbed by the body. Given healthy bone marrow and a normal supply of iron, people who are anemic because they have lost blood in an accident or in medical treatment can double or triple their production of red blood cells within seven to ten days.

The amount of iron actually supplied to our bodies, however, depends on the kind of food we eat. Fish and meat are rich in iron, and the iron in meat's hemoglobin is more easily absorbed than the iron in vegetables. However, many Asian populations are vegetarian because of religious tenets, and many of the world's poor are nearly vegetarian because they cannot afford meat products. Vegetables, with the exception of black beans, soybeans, and corn flour, contain little iron. Spinach is iron rich, but the oxalic acid it contains makes the iron unavailable to the body. Grain cereals, which the poor in Third World countries depend on for calories, contain phytates, calcium-magnesium complexes that inhibit the absorption of the small amount of nonheme iron in the grain. Bran and unpolished rice, in particular, contain large amounts of phytates. Bioengineers are trying to develop cereal crops that have less iron-inhibiting phytic acid.

Although staple foods are more available today for the poor in developing countries than in 1960, the availability of iron lags far behind. In those Asian populations where legumes are not a staple in the vegetable diet or in developing areas where people cannot afford meat, anemia is especially prevalent. Adding to the problem, the per capita consumption of legumes, a chief source of iron in these countries, has actually declined sharply.

Because it is highly unlikely that these people will begin to eat large amounts of meat and fish in the near future, the NAS sees nutritional education and bioengineering as their best hope. Various combinations of food can enhance or inhibit the body's absorption of iron, and eating meals with more iron-absorption enhancers and fewer inhibitors can make more iron available. In a meal with both meat iron and vegetable iron, the former improves the absorption of the latter. Iron-absorption enhancers include cooked cereals, fermented soy products, vitamin C, and germinated seeds. Studies show that drinking orange juice with an iron supplement can help iron absorption. Inhibitors include high-fiber foods, tea (including some herbal teas), coffee, and chocolate.

Unless meat is added to the diet, increasing iron intake can be difficult. For example, the average citizen of India would have to double or triple his caloric intake in order to increase his iron intake by 10–15 milligrams a day. As a result, WHO recommends fortifying basic foodstuffs with small amounts of iron. Studies have investigated the possibility of adding iron to curry in South Africa, fish sauce in parts of Thailand, maize flour in Venezuela, and table salt worldwide. For countries where iron deficiency is endemic, international nutrition groups have recommended fortifying food items with iron EDTA (sodium iron ethylenediaminetetraacetic acid). Most vitamin supplements contain iron in the form of ferrous sulfate. Whatever form of iron is used, a worldwide effort to eradicate iron deficiency will be an enormous and difficult undertaking.

Searching for a quick fix for iron deficiency in poor developing nations, physicians tried giving anemic infants large, prophylactic injections of iron. The results were disastrous. In 1970, intramuscular injections of iron to severely malnourished anemic children in South Africa caused a variety of fatal infections. When a New Zealand hospital gave iron shots to more than fifteen hundred at-risk Polynesian newborns during their first week of life in 1971 and 1972, the incidence of *E. coli* meningitis and septicemia jumped to ten or twenty times previous rates. Worse still, sixteen infants died. Apparently the extra iron benefited the *E. coli* bacteria more than it did the children.

Even the small amount of iron in a normal diet sickened severely malnourished people with parasitic infections. In 1970, in the midst of a six-year drought in the Sahel savannah of West Africa, crops and cattle were dying and famine was rife. Within five days of their arrival in a hospital in Niger, twenty-three patients and fifty-one of their relatives became sick with malaria. During dry weather, their malaria should have been quiescent. Hospital workers speculated that the reason for the outbreak might have been the hospital diet, which, though not ideal, was far more nutritious and iron rich than the famine sufferers had been accustomed to. Clinical and laboratory studies confirmed that iron in the hospital food had strengthened the *Plasmodium falciparum* parasites that cause malaria. Chelators, drugs that remove excess iron, ended the malarial attacks.

In 1977, M. John Murray, a medical school professor from the University of Minnesota, and his family were working in a refugee camp in the central African country of Niger. The camp was filled with four to six thousand Somali nomads, many of whom were iron deficient, not because of intestinal parasites but because their all-milk diet was low in iron. Murray, like most physicians, had been taught that iron deficiency made people susceptible to serious infections. But Murray was impressed to see that the iron-deficient nomads seemed to be quite free from infections. The Murrays conducted an experiment to find out why this was so. Randomly dividing the nomads into two groups, the Murrays gave one group 900 milligrams of ferrous sulfate and the other group a placebo. (Since the camp had only seven thousand iron tablets, they had to limit the number of treated patients to seventy-one.)

Observing the nomads closely for a month, the Murrays counted seven episodes of infectious disease in the placebo group, but among the nomads given iron there were thirty-six flare-ups of preexisting malaria, brucellosis, and tuberculosis. Brucellosis is an infectious bacterial disease that affects mostly cattle, pigs, and goats. Normally the Somali nomads, who raised animals and consumed milk as their major source of energy, were also prey to brucellosis. But the conditions were not normal: there had been no rain, no mosquitoes, and no overt attacks of malaria in the area for more than a year, and the nomads in the refugee camps had had no access to their animals or milk products for more than six months. The Murrays concluded that iron deficiency had suppressed some of the infections that the Somali nomads were constantly exposed to.

"Iron deficiency in Somali nomads may be part of an ecological compromise," the Murrays reported. "Iron deficiency, debilitating in some [nomads] but rarely fatal prevents the more serious consequences of potentially fatal infections with malaria, tuberculosis, and brucellosis." In fact, the Murrays warned, "It may be unwise to attempt to correct iron deficiency in the face of quiescent infection, especially in isolated societies where the natural ecological balance is often a first line of defense against severe infections."

Treating iron deficiency in the developing world was proving to be more complicated than in industrialized countries. But during the 1980s, the need to find a way to treat iron deficiency during early childhood became critical. In the past, physicians had assumed that the malaise characteristic of anemia was the result of low hemoglobin levels and the resulting poor oxygen transport. Then, in 1989, studies by two professors at the State University of New York in Syracuse alerted scientists to the fact that iron deficiency in children affects their intellectual development. The two were the late Frank A. Oski, an expert in children's blood disorders who later moved to Johns Hopkins Medical School, and his collaborator, Alice S. Honig, who was professor of child development.

By 1993, studies of infants, preschoolers, and schoolchildren were consistently reporting the same results: iron deficiency anemia was associated with poor performance in infant developmental tests, IQ tests, learning tasks, and educational achievement. The results were the same whether they came from the industrialized United States and Great Britain or from developing countries such as Guatemala, Thailand, and Indonesia.

"Few areas of developmental pediatrics observe such distinct patterns," noted Ernesto Pollitt, a Peruvian-born psychologist. There was even stronger and more consistent evidence for the cognitive damage caused by anemia than for lead poisoning, he said. In 1982, Pollitt, then at the Massachusetts Institute of Technology, demonstrated for the first time that even subclinical levels of mild iron deficiency could impede children's mental development. The prognosis for iron-deficient children worsened when Pollitt, now at the University of California at Davis, repeated his studies with iron-deficient and anemic preschool youngsters in Egypt and Guatemala. Those studies indicated that iron supplementation did not reverse all the damage to the children's development. Similar results have since been found in India, Papua New Guinea, Indonesia, and elsewhere in world. In 1985, Pollitt reported on a group of nine- to eleven-year-old Javanese children in rural Indonesia. Even after their anemia had been corrected, they still lagged behind well-nourished children.

The Javanese study left unanswered questions about the damage associated with milder iron deficiency anemia and about its effects in older children, whose brains were not developing at peak velocity. Pollitt thought that studying a larger group of children might reveal subtle distinctions that smaller studies could not. He also hoped he could bolster the Javanese study by replicating it in a different ecological setting, population, and cultural milieu.

He decided to carry out a new study in Thailand. Iron deficiency anemia there is a complex public health problem because of the interaction between intestinal parasites and the Thais' vegetable diet. Studies conducted in the 1990s in rural areas showed that 25 percent of the men were anemic, as were 45 percent of the women and children.

Compounding the problem was the high prevalence of hemoglobin disorders, especially thalassemia anemias, where iron treatment increases the risk of toxic iron overload. Moreover, many people in the developing world have both severe anemia and long-standing parasitic and bacterial infections. Approximately 90 percent of rural Thai schoolchildren had hookworm. Disease-causing microorganisms seem to compete for iron with the iron-binding molecules transferrin and ferritin. In people with too few transferrin molecules in the blood plasma to capture injected iron and keep it out of harm's way, pathogens can eat the iron and thrive.

Traveling to rural Thailand in the hopes of discovering the relationship between moderate iron deficiency and poor performance on school achievement tests, Pollitt and his colleagues divided 1,359 children aged nine to eleven years into three groups: those who had sufficient iron, those who were iron deficient, and those who were anemic. (At the time, this was the largest group of iron-deficient children to be studied.)

The children were given IQ and Thai language tests before and after a fourteen-week treatment with ferrous sulfate tablets or placebos of sweet cassava pills that looked like iron. Schoolteachers distributed the pills without knowing the iron status of the children or contents of the tablets. At the end of the study, the group that began with healthy iron levels still had significantly higher scores on the IQ and language tests than either the anemic or iron-deficient children. Despite treatment, the iron-deficient children did not show the expected improvement in their educational scores. Iron treatment for fourteen weeks could not correct the difference. When Oski conducted his studies in 1989, he had expected that the effects of iron deficiency would turn out to be reversible, but Pollitt showed that they were not. Not only can iron deficiency affect children's development, but much of the damage follows them throughout their lives.

Since Pollitt's study, other researchers have also found irreparable damage in iron-deficient youngsters in double-blind studies conducted in Costa Rica, Chile, and parts of Guatemala. For example, scientists tested 191 infants for iron deficiency in an urban community near San José, the capital of Costa Rica, during the mid-1980s. Several indicators were used: venous levels of hemoglobin, transferrin saturation, serum ferritin, and erythrocyte protoporphyrin. Five years later, Betsy Lozoff of the University of Michigan, E. Jimenez, and A. W. Wolf retested 163 of the original 191 children. By then, all the five-year-olds had normal iron levels, but those who had had moderately severe anemia as infants scored lower in mental and motor functioning than the other children, even when socioeconomic factors were considered. Lozoff concluded in a 1991 *New England Journal of Medicine* article that infants with iron deficiency anemia risk long-lasting developmental damage. She advocated vigorous efforts to prevent iron deficiency.

However, she realized that no one had yet proved that iron deficiency was actually responsible for the children's lower test scores in infancy and at five years of age. She decided to carry out a new study, which involved testing sleeping infants and children for auditory brain-stem responses. Such measurements, which are an indication of the health of the central nervous system, had not been conducted on iron-deficient children before. The presence of a highly sophisticated infant neurophysiology laboratory in Chile made the experiment possible.

Chile is a democracy with a highly literate population and comprehensive health care, and infant health is generally excellent there. Malnutrition, hookworm, high lead levels, and hemoglobin disorders are virtually absent. Iron deficiency, however, is common because the diet is low in iron. Lozoff traveled to Chile, where she, Dr. Manuel Roncagliolo, and other neurophysiologists studied the brain-wave patterns of fifty-five sleeping children.

She used a noninvasive test that analyzed a particular electrical wave as it traveled from an acoustic nerve to the brain. In early childhood, the auditory system, part of the central nervous system, is still developing and maturing. Studying the fifty-five sleeping children, the researchers found that children who had been anemic at six months had different brain-wave patterns than those who had had normal iron levels. And despite effective iron therapy, the differences became more pronounced at twelve and eighteen months of age. Lozoff concluded that iron deficiency anemia in infants may indeed be associated with central nervous system damage. Such differences might indicate developmental problems for the fatty sheaths that cover nerve fibers and for language learning.

Over the last decades, while nutritionists and psychologists were documenting the damage caused to children by iron deficiency, other scientists have been studying iron metabolism from a molecular point of view. With breathtaking speed, they have succeeded in expanding our understanding of how iron functions within the human body. Because of their work, we have been able to follow life's legacy of iron full circle, from the formation of our iron-based planet to the human occupation of it.

Iron formed Earth's core and almost certainly provided much of the energy needed for early life. But today, confronted with living creatures that luxuriate in deep-sea volcanic vent communities permeated with iron compounds, some scientists think that life may have originated on iron-sulfur surfaces. With the arrival of atmospheric oxygen, the iron that early life forms depended on rusted and became unavailable to them. To walk life's tightrope between iron deficiency and iron toxicity, primitive organisms developed molecular tools that seize biologically scarce iron, store it until needed, and shuttle it around our bodies. Some mud-dwelling bacteria built tiny magnets in order to reach their preferred oxygen level, while some more-complex animals evolved to migrate enormous distances by sensing Earth's iron-based magnetic fields. The acme of these iron-dependent molecules is surely the hemoglobin and myoglobin that transport oxygen and dump it into the tissues of energy-hungry vertebrates, including people. From these iron-based molecules, the metal journeys through the biological world from a molecular scale of life to a global scale as iron fertilizes oceans, soils, and plants.

Despite all the progress made over the past twenty years in understanding iron, our knowledge of its role in human health is far from complete. Some day an understanding of iron biochemistry could help eliminate widespread diseases, such as the thalassemias and hemochromatosis. Learning how to use the extraordinarily iron-replete process of nitrogen fixation could help us protect millions of youngsters around the world from the ravages of dietary deficiencies.

But human health depends on more that just our scientific understanding of iron's role in biology. It also depends on how humans view their relationship with the other iron-dependent systems of our planet: iron atoms link us with virtually every other living creature on Earth, and with the plants and soils and oceans upon it.

Thus, the legacy of our early iron-based biochemistry presents us with a challenge. We must continue to decipher iron's secrets, legacies from our ancient biochemical past, when oxygen was scarce and iron was more precious than gold. Doing so may help us explain the origin of life on Earth. But even more important, it may help us to become better stewards of life on our planet.

GLOSSARY

Adult hemoglobin. The hemoglobin of normal adults, which has two alpha chains and two beta chains.

Aerobic. With oxygen.

Alloy. A mixture of a metal and one or more other elements, generally another metal.

Alpha chain. One of the two most common hemoglobin subunits, the other being the beta chain. Alpha chains contain 141 amino acids, of which 38 differ from those in beta chains. Alpha chains combine with beta chains to make normal adult hemoglobin and with gamma chains to make fetal hemoglobin.

Alveolar. Of or pertaining to air sacs in the lungs.

Amino acid. An organic molecule possessing an amine group ($R-NH_3$) and a carboxylic acid group ($R-COOH$). Amino acids are the building blocks of all peptides and proteins. There are twenty common amino acids, nine of which, termed essential, cannot be synthesized by vertebrates and must be obtained from the diet.

Anaerobic. Without oxygen.

Anemia. A medical condition in which the blood has an abnormally low oxygen-carrying capacity. In several blood diseases, the oxidation (the loss of an electron) of iron atoms in hemoglobin inside red blood cells quickly destroys the red blood cells and prevents them from transporting oxygen through the body. Anemia can be genetic in origin, or it can be caused by toxic substances, such as aniline dyes or nitrate-contaminated water, which break the membranes of the blood cells.

Angstrom. A unit of length used to measure the size of atoms. An angstrom is 10^{-10} meters long. An atom measures approximately 2 to 4 angstroms across. Angstrom is symbolized by Å.

Archaea. One of the three branches on Carl Woese's tree of life. Formerly called archaeabacteria, these primitive one-celled organisms were among the first life on Earth. The other branches on the tree of life are bacteria (procarya) and eucarya, the latter branch including fungi, plants, and animals.

Arthropoda. The largest phylum in the animal kingdom. It includes invertebrates such as horseshoe crabs.

Asteroid. A small planetary body, one of many that revolve around the sun, mainly between the orbits of Mars and Jupiter. Asteroids are bigger than meteorites.

Atom. The basic building block of chemistry. An atom is the smallest unit of an element that has the characteristic chemical properties of that element.

ATP (adenosine triphosphate). The carrier of chemical energy in living organisms.

Autolithotrophic organism. A microbe able to live on rock.

Bacteria. Primitive cells without nuclei (formerly classified as procaryotes). The term is also sometimes used as a general term for microorganisms that are not classified as fungi.

Bacteroids. Compartments created by symbiotic bacteria inside the nodules of leguminous plants. These are the spaces in which nitrogen fixation occurs. See figure 7.4 and *nitrogen fixation.*

Banded iron formations. Intensely red iron-ore deposits formed when ferric oxide precipitated from ocean water beginning about 2.7 billion years ago.

Beta chain. One of the two most common hemoglobin subunits, the other being the alpha chain. Beta chains are made of 145 amino acids, of which 38 differ from those in the alpha chain.

Biomineralization. The process that turns animal or plant matter into mineral material.

Biomass. The total weight or body mass of all organisms in a given volume.

Carbohydrates. A large group of compounds that contain carbon, hydrogen, and oxygen. Sugars and starches are carbohydrates.

Carbon monoxide (CO). Carbon monoxide normally occupies about 1 percent of the iron sites in healthy hemoglobin, but smokers who inhale add up to 15 percent more carbon monoxide to their hemoglobin. Smokers who donate blood are often told to wait at least one hour before lighting up; giving blood lowers the body's supply of hemoglobin, so carbon monoxide would fill an even higher percentage of the iron sites in the remaining hemoglobin. In heavily trafficked enclosed parking garages and tunnels, and in homes with inefficient and unvented stoves, the air often contains 100 parts per million of carbon monoxide. At that level, 16 percent of hemoglobin's iron sites fill with carbon monoxide instead of with oxygen, causing headaches and shortness of breath. Carbon monoxide packs a one-two punch. Over all, when it binds to 40 percent of the hemoglobin, it lessens the oxygen-carrying capacity of the blood. Death occurs when carbon monoxide prevents 50 percent of the iron sites from accepting oxygen.

Carrier. An organism that is infected with a pathogen but shows no symptoms or only mild symptoms. In the case of an inherited genetic disease, the carrier has one normal and one abnormal version of a gene; two abnormal genes must be inherited for the disease to occur.

Catalyst. A substance that speeds up a chemical reaction without being permanently changed.

Cell. The basic structural and functional unit of all living organisms.

Charge. The electrical charge on an object. It may be positive, negative, or zero.

Chelator. A metal-grabbing complex used to remove metal, such as iron or lead, from the body. The term *siderophore* was coined to describe the strong iron-binding chelators that bacteria manufacture.

Chemical bond. An attractive force that holds atoms together in a compound.

Chemical reaction. The process that changes (one or more) elements or compounds into new compounds.

Chloroplast. The site of photosynthesis within cells.

Cobalt. A metallic element with twenty-seven protons in its nucleus. One of the five ferromagnetic elements, it is less magnetic than iron. Its symbol is Co. See *exchange interaction.*

Conformational motions. When a molecule such as hemoglobin changes shape, its function changes too.

Cooley's anemia. A genetic disease in which the hemoglobin has too few beta chains and the unpaired alpha chains are highly toxic. Cooley's anemia is one of the most destructive forms of thalassemia.

Corynebacterium diphtheriae. The bacterium that causes diphtheria.

Crystallography. The use of x-rays to determine the arrangement of atoms within a crystal, that is, within solids whose atoms are arranged in a regular and repeated pattern.

Cyanobacteria. One-celled organisms formerly classified as blue-green algae. These bacteria use light for energy and iron-sulfur proteins to produce organic molecules and oxygen from water and carbon dioxide.

Cytochrome oxidase. A molecule with an iron and copper center. It is the enzyme responsible for 90 percent of the cell's uptake of molecular oxygen.

Deep divers. Marine mammals who can dive to depths of three-tenths of a mile (500 meters) or more. These mammals store a large percentage of their oxygen in myoglobin, the blood in muscles. For example, a Weddell seal stores only 5 percent of its oxygen in its lungs but 66 percent in its blood and 29 percent in its muscle, for a total of 95 percent outside its lungs. In comparison, people hold 24 percent of their total oxygen in their lungs, 54 percent in their blood, and only 15 percent in muscle. Among other marine creatures, some whale species have enough myoglobin in their muscles to carry half again as much oxygen as their red blood cells can accommodate. Birds can also store oxygen in their myoglobin. The emperor penguin, which can dive one-third of a mile deep, can store 47 percent of its overall body oxygen in muscle.

Desferal. The trade name for deferoxamine B mesylate, a naturally occurring siderophore produced by the bacteria *Streptomyces pilosus.* Desferal, which came on the market in 1962, is used to treat toxic iron overload. Besides treating thalassemias, it is used in other anemias in which high iron absorption leads to toxic iron overloads. Deferoxamine rarely causes toxicity or allergies, but it can be highly irritating to the skin.

Deoxyribonucleic acid (DNA). The genetic material of most living organisms, the carrier of genetic information. Through the transfer of genetic information to RNA (ribonucleic acid), DNA ultimately controls protein synthesis in cells.

Diatom. Bacillariophyta, one-celled algae found in plankton.

Dip compass. A compass that measures the inclination of the earth's magnetic field and can thus be used as an indication of latitude; also called an inclination compass.

Domain walls (boundaries). Areas between magnetic domains, in which unpaired electron spins are not aligned.

Dysprosium. A metallic rare-earth element with sixty-six protons in its nucleus. One of the five ferromagnetic elements, it is symbolized by Dy. See *exchange interaction.*

Electric forces. Forces produced by electric charges at rest. These forces are one aspect of electromagnetism, the science of charge and of the forces and fields associated with charge.

Electromagnetism. The science of charge and of the forces and fields associated with both static and moving charge.

Embryo. A vertebrate before it is born or hatched; in humans, the result of conception up to the third month of pregnancy.

Embryonic hemoglobin. The embryo's hemoglobin for the first eight weeks after conception. It has two zeta chains and two epsilon chains.

Endocytosis. The process by which mammalian cells assimilate large particles, including the iron in transferrin. See figure 3.6.

Enterobactin. The world's most powerful iron-grabbing siderophore, made by the intestinal bacterium *Escherichia coli (E. coli).*

Enzyme. A protein that catalyzes, or facilitates, a biochemical reaction. Enzymes are often named by adding the suffix "-ase" to the name of the reaction they catalyze. For example see *hydrogenase, cytochrome oxidase, nitrogenase,* and *peroxidase.*

Erythrocytes. Red blood cells.

Erythropoietin (EPO). A hormone produced in the kidney and liver. It stimulates the bone marrow to produce red blood cells. It is used to treat anemia caused by kidney disease, AIDS, and cancer.

Escherichia coli. Commonly called *E. coli,* an intestinal bacterium that makes the world's most powerful siderophore, enterobactin. Under the right growth conditions, a culture of *E. coli* can double the number of its cells every twenty minutes.

Eucaryote. A unicellular organism having a nucleus (see also *procaryote*); also called eubacteria and, more recently, eucarya.

Exchange interaction. The interaction responsible for ferromagnetism. The exchange interaction is a special quantum effect in which the unpaired spins of electrons of one atom force spins on a neighboring atom to align. The characteristic of ferromagnetic materials is the presence in the material of a large internal magnetic field. Only five elements are ferromagnetic: iron, cobalt, nickel, gadolinium and dysprosium. Of the five, only the first three are ferromagnetic at room temperature, 70 degrees F (20 degrees C); gadolinium and dysprosium must be cooled.

Exponent. A number printed in small type above and to the right of another number to indicate the number of times the latter must be multiplied by itself. Thus 10^8 means that 10 must be multiplied by itself eight times. The number 10^{-8}, which has a negative exponent, means that 10 must be divided by 100,000,000.

Ferric iron (Fe^{+++}). An iron atom that has lost three of its twenty-six electrons. This chemical state is called the oxidized form of iron. See figure 1.1.

Ferrihemoglobin. A brown pigment formed when hemoglobin is released from the red blood cells and exposed to the atmosphere; also called methemoglobin.

Ferritin. A large (about 120 angstroms), hollow protein shell built to store iron atoms. Ferritins occur naturally in all cells, including those in plants. In mammals, ferritin iron-storage proteins are found in the cells of the liver, spleen, muscles, and blood serum.

Ferromagnetism. The property exhibited by materials that attract or repel iron. A magnetic field is naturally created around a ferromagnetic solid. The characteristic of ferromagnetic materials is the presence in the material of a large internal magnetic field. Only five elements are ferromagnetic: iron, cobalt, nickel, gadolinium, and dysprosium. Of the five only the first three are ferromagnetic at room temperature; gadolinium and dysprosium must be cooled. See chapter 4 and *exchange interaction.*

Ferrous iron (Fe^{++}). An iron atom that has lost two of its twenty-six electrons. This chemical state is called the reduced form of iron. See figure 1.1.

Fetal hemoglobin. A transitional form of hemoglobin made by the fetus; also called gamma hemoglobin. It consists of one pair of gamma chains and one pair of alpha chains. By the time the fetus is born, 50 percent of its chains are adult alpha chains, 30 percent are fetal gamma chains, and only 20 percent are adult beta chains. An infant does not have fully adult hemoglobin—with equal numbers of alpha and beta chain pairs and a few remaining gamma chains—until twenty-six weeks after birth. Fetal hemoglobin has a 30 percent higher affinity for oxygen than the mother's own hemoglobin (see figure 5.8). It binds oxygen more tightly than adult hemoglobin or myoglobin because it binds less 2,3-DPG.

Fetus. Embryo of a mammal. A human embryo is called a fetus when it is between eight weeks of age and birth.

Field. The region of space in which one body can exert a force on another body. The objects do not have to touch. The most obvious example is the attractive gravitational force exerted by the sun on the planets. The force of an electrical or magnetic field can be either attractive or repulsive. Like charges and like magnetic poles repel each other; unlike charges and unlike magnetic poles attract each other.

Fission. The breaking apart of an atomic nucleus into lighter parts.

Flagella. Whip-like arms that allow bacteria to move. Flagella are designed to rotate, propelling the bacteria into regions that are more physically or chemically favorable to the bacteria.

Fool's gold. A nonmagnetic iron-sulfur complex known as pyrite (FeS_2).

Free iron. Iron atoms that are not stored in ferritin, transferrin, hemoglobin, or lactoferrin. They can catalyze dangerous free radical reactions in cells. See *free radical.*

Free radical. A very reactive atom or molecular fragment with one unpaired electron. Free radicals rob other atoms or molecules of their electrons, thereby damaging cells. They are considered to be a major initiator of cancerous growth.

Fusion. The combination of two or more light atomic nuclei to form a heavier nucleus.

Gadolinium. A ferromagnetic magnetic element, a rare-earth metal with sixty-four protons in its nucleus. Its symbol is Gd. See *exchange interaction.*

Gamma rays. High-energy radiation emitted by atomic nuclei. Wavelengths of gamma rays are in the range of 10^{-10} to 10^{-18} meters.

Gauss. The unit of measurement for magnetic flux density. It is symbolized by G and equals 10^{-4} tesla (10,000 gauss equals 1 tesla). Earth's magnetic field is about half a gauss.

Geobacter metallireducens. An anaerobic bacteria that imports ferric iron and expires ferrous iron. Either on the cell surface or very close to the cell surface, the expired iron participates in a reaction which produces magnetite. By weight, GS-15 can produce five thousand times more magnetite than an equivalent biomass of *Magnetospirillum magnetotacticum.*

Globin. The protein chain component of the hemoglobin molecule.

Greigite. A magnetic iron-sulfur compound (Fe_3S_4).

Heme. The protoporphyrin ring, composed of carbon, nitrogen, and hydrogen atoms, around a single iron atom that binds the oxygen in hemoglobin, myoglobin, and cytochromes.

Hemochromatosis. The most common genetic disorder in the United States. It makes the body absorb too much iron from food. Hereditary hemochromatosis is caused by one defective gene that encodes an iron regulatory protein.

Hemocyanin. The "hemoglobin" found in horseshoe crabs, scorpions, spiders, and garden and sea snails. Like hemoglobin, it is a reversible oxygenator. But at its active center, the site where the oxygen binds, iron is absent. Instead, the oxygen is held by two copper atoms.

Hemoglobin. The respiratory pigment in red blood cells. Each of hemoglobin's four chains is composed of approximately 2,000 atoms: 1,000 hydrogens, 700 carbons, 200 nitrogens, 200 oxygens, 2 sulfurs, and 1 iron atom. Thus, with a molecular weight of 64,450, hemoglobin weighs approximately 64,000 times more than a hydrogen atom. The term deoxygenated hemoglobin refers to a hemoglobin molecule where the iron sites have no molecules of oxygen attached.

HNLC. High nutrient–low chlorophyll regions of the oceans' surfaces that are rich in major nutrients but quite infertile.

Hydrogenase. A family of iron-containing enzymes that catalyze the conversion of two hydrogen ions (H^+) to molecular hydrogen (H_2). An important structural feature of this enzyme is its iron-sulfur clusters.

Hydrops fetalis. A thalassemia disease in which a baby is born with no alpha hemoglobin chains.

Hydrous. Containing water.

Inclination compass. See dip compass.

Inorganic. Not made of living or formerly living things.

Iron EDTA. Sodium iron ethylenediaminetetraacetic acid.

Iron-sulfur molecules. Clusters composed of one to eight iron atoms attached to one or more sulfur atoms. They are ubiquitous in all living organisms including the most ancient bacteria. They are star performers in electron transfer reactions, and they stabilize protein elements and serve as regulatory sensors of iron and molecular oxygen.

Isotope. Any of two or more forms of an element with the same number of protons but a different number of neutrons in its atomic nucleus. All elements have isotopes, and all the isotopes of a particular element have the same or quite similar chemical characteristics.

Lampshells (brachiopods). Bivalved marine worms resembling Roman oil lamps. They are found in fossil records dating back to 500 million years ago.

Leghemoglobin. The only hemoglobin known in plants. Structurally similar to myoglobin. Its function is to bind oxygen (O_2).

Line of force. A line in a diagram that shows the direction of a magnetic or electrical force field.

Magnetic bacteria. All the species of bacteria which manufacture an internal compass of a magnetic material. They use this compass to align themselves along the earth's magnetic field lines. They travel along these lines finding the place in the mud where the amount of oxygen perfectly matches their respiratory requirements. The cells of some of these stalk-like bacteria are coated with a hydrous iron oxide $Fe(OH)_3$. These iron oxide sheaths give the bacteria ballast and allow them to sink below the water's oxygen-rich surface. This eases their problem of locating the correct oxygen level.

Magnetic domain. In a ferromagnetic solid, a region in which the unpaired spins of neighboring electrons align. The size of the domain depends on the constituents of the solid and on temperature. As the domain becomes larger, it becomes more difficult to make all these electron spins point in the same direction. When the domain size gets too large, a complex force compels electrons to reorient themselves in a different direction, splitting off as a second domain. Large and powerful magnets are assembled from many single domains, with all their odd electrons pointing in the same direction. The magnet's force depends on the number of atoms with identically pointing electrons, so doubling the number of electrons that point in one direction makes the magnet twice as strong.

The term "single magnetic domain" refers to the smallest possible amount of material that can exhibit ferromagnetism. Each single domain measures about 150 to 200 angstroms across and contains several million iron oxide molecules. Chapter 4 describes magnetic bacteria, such as *Magnetospirillum magnetotacticum,* that manufacture single magnetic domains and string twenty to thirty of them together to form an internal compass. To fit inside a bacterium such as *Magnetospirillum magnetotacticum,* the magnets had to be less than four-thousandths of an inch (about one-millionth of a meter) across; so the bacterium "discovered" the size of a single magnetic domain. Then it lined these domains up in a chain to overcome thermal jostling by adjacent molecules. This magnetic material accounts for 1.5 percent of the bacterium's dry weight, an enormous amount for a microorganism. The glossary entry *ferritin* is also related to magnetism.

Magnetic equator. A line running close to the geographic equator, where Earth's magnetic field lines are parallel to Earth's surface. The magnetic equator passes through Brazil, Africa, India, and Indonesia.

Magnetic field. The force-filled region around a magnetic body. See *field.*

Magnetism. A phenomenon characterized by fields of force produced by moving charges. There are three kinds of magnetism: ferromagnetism, paramagnetism, and diamagnetism. The familiar magnetism of iron bar magnets is a type of ferromagnetism. Both ferromagnetism and its much weaker sister paramagnetism are caused by the electron's spin. In diamagnetism, the motion that creates the magnetic effect is the orbital motion of the electrons around the nucleus. Magnetism is one aspect of electromagnetism, the science of the forces and fields associated with static and moving charges.

Magnetite. A magnetic crystal of ferrous and ferric oxides ($FeO + Fe_2O_3$); also called lodestone.

Metabolism. (1) The breakdown of food into substances and energy that a living organism can use. (2) The physical and chemical processes that sustain life.

Meteorites. Chunks of matter, made of rocks or iron, from interplanetary space that enter Earth's atmosphere and strike Earth's surface.

Methemoglobin. A brown pigment formed when hemoglobin is released from a red blood cell and exposed to oxygen; also called ferrihemoglobin.

Micro. Generically a prefix meaning very small. It also has an exact arithmetic meaning: 10^{-6}, that is, 1 divided by 1,000,000.

Microorganism. A microscopic bacterium, archaea, protozoa, yeast, virus, algae, or fungus.

Mitochondria. The minute cellular biochemical furnaces, shaped like filaments or rods, that provide energy to living cells.

Mössbauer spectroscopy. An experimental technique, named for the 1961 Nobel Prize–winning physicist Rudolph Mössbauer, that is especially suited for the study of molecules that contain iron (see chapter 4).

Mutation. A sudden change in the genetic material of a cell that changes the characteristics of the cell and its subsequent generations.

Myoglobin. The respiratory pigment found in muscle cells. Myoglobin consists of a single amino acid chain folded around a heme and is a primitive version of hemoglobin. See chapter 5 and figure 5.1 for a complete description of its function.

Neutron. A particle inside the nucleus of an atom. A neutron has no electric charge. Its mass is about the same as a proton. The mass of a neutron is about two thousand times that of an electron.

Nickel. The twenty-eighth element on the periodic table. A metal, it has twenty-eight electrons. Its symbol is Ni. One the five ferromagnetic magnetic elements, nickel is less magnetic than iron or cobalt.

Nitrogen fixation. The conversion of nitrogen molecules from the atmosphere into nitrogen atoms, with the subsequent formation of ammonia and release of hydrogen: $N_2 + 8H^+ + 8e^- \rightarrow 2NH_3 + H_2$. With additional reactions plants incorporate the ammonia as an amide, an ammonia molecule with an extra electron. Nitrogen fixation is the first step in the synthesis of an amino acid.

Nitrogenase. A large iron-sulfur protein that catalyzes nitrogen fixation in legumes and other organisms.

Nod gene. A nodule-making gene in bacteria that disables a legume's defenses so the bacteria can enter the plant and fix nitrogen.

Nuclear force. A strong, short-range force that holds together the protons and neutrons of atomic nuclei.

Nucleic acid. A large macromolecule composed of nucleotides. DNA and RNA are nucleic acids.

Nucleotide. A compound composed of a base (derived from purine, pyrimidine, or pyridine), a five-carbon sugar, and a phosphoric acid group. Nucleotides are the components of nucleic acids. Nucleotides are short snippets of nucleic acid.

Organic. Made of living or formerly living things.

Organic matter. Carbohydrates. A large group of compounds which contain carbon, hydrogen, and oxygen. Sugars and starches are carbohydrates.

Organism. An individual living being, whether a plant, animal, bacteria, archaea, or other form.

Oxidation. A chemical reaction in which an atom loses at least one electron. Iron's oxidized state (Fe^{+++}) is called ferric iron.

Oxyhemoglobin. Fully oxygenated hemoglobin. All four of hemoglobin's iron atoms bind an O_2 molecule.

Ozone (O_3). A molecule in which three oxygen atoms are bound together.

Pathogenic. Disease-causing.

Peanut worm (sipunculid). A rubbery unsegmented marine worm, often spindle shaped and with a tentacled head.

Peptide. A short chain of amino acids which are linked by bonding the carboxyl (CO_2) end of one acid to the amine (H_3N) end of the next. Proteins are peptides that are longer than fifty amino acids.

Peroxidase. An iron-containing enzyme that promotes the conversion of hydrogen peroxide to water.

pH. A measure inversely related to the concentration of hydrogen ions (H^+) in a system. One of the first concepts introduced in high school chemistry is the comparison between acids, such as hydrochloric acid, and bases, such as soda or potash: an acid is a chemical that can shed protons and is thus a source of hydrogen ions; a base is a chemical that can accept protons. A system with a relatively few hydrogen ions has a high pH and is said to be basic, or alkaline; and a system with many hydrogen ions has a low pH and is called acidic. The pH of a solution of pure water at room temperature is 7.0. The chemistry of living systems depends on a relatively stable pH, nominally pH 7. For example, normal blood pH is 7.4.

Photosynthesis. A chemical process in which the cells of living organisms use energy from sunlight to convert water and carbon dioxide into organic carbon compounds and molecules of oxygen. Living organisms that use energy from the sun are called phototrophs. Some phototrophs, such as plants and certain species of bacteria, use chlorophyll molecules to extract energy from sunlight. Not all phototrophs have chlorophyll, however. Some have a different light-activated energy transfer molecule called rhodopsin, and some microbial evolutionists think that rhodopsin evolved before chlorophyll.

Phyto. Prefix: Having to do with plants; for example, phytoplankton and phytoferritin.

Phytoplankton. Microscopic floating plants in oceans. They form the bottom of the marine food chain.

Phytosiderophores. Siderophores made by plants.

Plankton. Microscopic plants and animals floating on the oceans' surface.

Polymerization. As used in this book, the process in sickle cell disease by which the hemoglobin molecules in red blood cells stick together to form long, stiff rods.

Porphyrin ring. The ring of carbon, nitrogen, and hydrogen atoms that surrounds an iron atom in hemoglobin, myoglobin, and cytochromes.

Procaryote. A unicellular organism having no nucleus. Classified as bacteria on Woese's tree of life. See also *eucaryote*.

Protein. A large, complex molecule made up of amino acids. Cellular proteins carry on enzymatic activity, that is, they accelerate specific chemical reactions in cells. Proteins are responsible for cellular metabolism. They are also the organic basis for many structures such as hair, muscle, skin, and so on.

Protoporphyrin. The organic part of the heme.

Pyrite. A nonmagnetic iron-sulfur complex known as fool's gold (FeS_2).

Radiation. The emission and propagation of energy-transmitting mechanical or electromagnetic waves, for example, sound and light.

Recombinant DNA technology. The manipulation of DNA molecules by a variety of biochemical techniques. Specific regions of the molecule may be cleaved, spliced, copied, sequenced, and altered, allowing researchers to study the functions of proteins.

Reduction. The chemical reaction that occurs when an ion gains an electron as it forms a compound with another atom. Ferrous (Fe^{++}) is called the reduced state of iron; the iron has lost two of its electrons. If you are not a chemist, the use of the term is curious because "reduced" is used here as a comparison with the "oxidized" state of iron, the ferric state (Fe^{+++}), the state in which the iron has lost three of its electrons. See also *oxidation.*

Respiration. Various biochemical reactions in which absorbed chemicals—from food, for example—are broken down into simpler compounds with the release of energy. Aerobic respiration involves oxygen and produces water and carbon dioxide; anaerobic respiration does not.

Reversible oxygenation. Hemoglobin's binding of oxygen in the lungs and dumping of oxygen in the tissues is an example of reversible oxygenation.

Rhizobium bacteria. Bacteria that form a symbiotic relationship with legumes to fix nitrogen in the plant's roots.

Rhodopsin. A light-activated energy transfer molecule similar to chlorophyll.

Rhodotorula pilimanae. A fungus that makes an enormous number of siderophores in iron-deficient soils. This fungus is a veritable sorcerer's apprentice; when grown in an iron-deficient medium, it produces half an ounce (10 grams) of siderophores per quart (liter) of cellular matter. This a staggering amount for a microorganism.

Ribonucleic acid (RNA). A molecule that transfers information from DNA to the cells or organs. It is responsible for the synthesis of enzymes.

Rust. Iron oxide, a compound of ferric iron (Fe^{+++}) and oxygen.

Rusty liver disease. A generic term applied to the pathological problem of a nonfunctioning, iron-overloaded liver.

Siderophores. Iron-grabbing molecules that make inorganic iron, such as ferric oxide, available for biological use in microorganisms and plants. Siderophores have several different structures, each a minor variation on a three-armed theme. See figure 3.1 for the structure of enterobactin, a typical siderophore. The iron is bound in the central core of the molecule. Oxygen atoms in the siderophore attach themselves to all six of the iron atom's bond, coordination sites. The central iron ion acts as a positive center. The negative organic part of the molecule, the black base shown in figure 3.1, donates electrons to the metal ion. The result is a so-called coordination compound, a hybrid mix, part organic and part inorganic. Siderophores can be used to treat patients suffering from iron overload. See *Desferal.* Other metal-grabbing molecules—whose structures are similar to siderophores—called chelators are used to treat lead and copper poisoning.

Species. A group of living organisms with a common ancestry and the ability to breed together.

Spleen. The organ that cleans up dead red blood cells. The spleen contains large numbers of ferritin molecules.

Stromatolites. Fossilized mounds, probably made by microorganisms such as cyanobacteria starting about 3.5 billion years ago. Stromatolites are found in Yellowstone National Park and in Western Australia.

Supernova. A star that explodes, suddenly giving off 10^{10} times more energy in the form of light.

Superparamagnetism. A kind of weak magnetic system in which the spins of some of the electrons in iron oxide microcrystals are aligned with each other but as a group swivel freely and point in the same direction as the prevailing magnetic field.

Symbiotic relationship. A usually beneficial relationship between members of two species.

Tesla. The unit of measurement for magnetic flux density; 1 tesla equals 10,000 gauss. One nanotesla equals 10^{-9} tesla.

Thalassemias. A group of inherited, incurable diseases of hemoglobin that occur where malaria is endemic. They are marked by an overproduction of alpha or beta chains.

Thermal energy. Random molecular motion caused by heat energy.

Thermodynamics. The science of energy transfer.

Ton B. A protein inside cells that helps siderophores enter the cytoplasm of cells.

Trace element. An element that is present in very small quantities.

Transferrin. A molecule, composed of a single chain of amino acids, that transports iron atoms in the body. The amino acid chain of the transferrin molecule is organized into two very similar lobes, each of which contains an iron-binding site, so that each transferrin molecule can carry two iron atoms to various parts of the body. Eighty percent of the transferrin bound iron in the blood plasma heads for the bone marrow for the production of red blood cells. Transferrin also carries aging red blood cells and iron absorbed from food. Transferrin delivers iron to the liver for the synthesis of other iron complexes and for storage in ferritin. Transferrin molecules come in several varieties, depending on where they work. Blood serum transferrin carries iron from the ferritin storage cages to all cells, especially to developing red blood cells in the bone marrow. Lactoferrin—found in milk, tears, saliva, pancreatic juices, and mucus—may help infants fight bacterial infections. Lactoferrin binds iron atoms even more tightly than serum transferrin does. Since many disease-causing bacteria thrive on iron, lactoferrin thwarts their growth by reducing the amount of iron available. Other transferrins are found in bird egg white, ovotransferrin, and in human tumors, melanotransferrin. See *ferritin*.

Trilobites. A class of extinct marine arthropods from the Paleozoic era.

Ultraviolet rays. Electromagnetic radiation with wavelengths between 4 and 400 nanometers and between visible light and x-rays.

Units of measurement.

Length

1 kilometer = 1,000 meters = 0.621 mile = 5,280 feet.

1 meter = 39.37 inches.

1 centimeter = one-hundredth of a meter = 0.01 meter
= 10^{-2} meter = 0.39 inch.

1 inch = 2.54 centimeters.

1 millimeter = one-thousandth of a meter = 0.001 meter
= 10^{-3} meter.

1 micron = 1 micrometer = one-millionth of a meter
= 0.000001 meter = 10^{-6} meter.

1 nanometer = one-trillionth of a meter = 0.000000001 meter
= 10^{-9} meter.

1 angstrom = 0.0000000001 meter = 10^{-10} meter.

Volume

1 liter = 1.057 quarts.

1 milliliter = one-thousandth of a liter = 0.001 liter = 10^{-3} liter.

Mass with weight equivalents

1 kilogram is equivalent to a weight of 2.205 pounds on Earth.

1 gram is equivalent to a weight of 0.035 ounce on Earth.

1 milligram = one-thousandth of a gram = 0.001 gram
= 10^{-3} gram.

Pressure

1 mm Hg = 0.0013 atmosphere (the equivalence between the English system, lbs/in^2, and the metric system is often remembered as 1 atmosphere = 14.7 lbs/in^2 = 760 mm Hg).

Valence. A number that describes the ability of an atom to react or combine with other substances in terms of the number of hydrogen atoms that the element could bind with. Hydrogen, the element with one electron, has a valence of one. For example, the symbol Fe^{+++} means that the iron atom has lost three of its electrons. Iron has twenty-six protons, so iron in this state has three remaining positive units of electronic charge. This directs its binding to other ions during the formation of compounds. Chemists characterize many subtleties of bonding in terms of electron transfer. There are two extremes: in ionic bonding, the electron from one atom is wholly transferred to another atom, whereas in covalent bonding, electrons are shared between the atoms.

Vertebrate. Any animal with a backbone.

X-rays. High-energy electromagnetic waves between 0.05 and 100 angstroms long. They can penetrate most matter.

Zooplankton. Microscopic floating animals that eat microscopic floating plants in the oceans.

BIBLIOGRAPHY

General References

Biddle, W. *A Field Guide to Germs*. New York: Henry Holt, 1995.

Isselbacher, K. J., E. Braunwald, J. D. Wilson, J. B. Martin, A. S. Fauci, and D. L. Kasper, eds. *Harrison's Principles of Internal Medicine*. 14th ed. New York: McGraw-Hill, 1998.

Schopf, J. W., ed. *Earth's Earliest Biosphere: Its Origin and Evolution*. Princeton, N.J.: Princeton University Press, 1983.

Smith, P. J., ed. *The Earth*. New York: Macmillan, 1986.

Stryer, L. *Molecular Design of Life*. New York: W. H. Freeman, 1989.

Verschuur, G. L. *Hidden Attraction: The Mystery and History of Magnetism*. New York: Oxford University Press, 1993.

Weatherall, D. J., J.G.G. Ledingham, and D. A. Warrell, eds. *Oxford Textbook of Medicine*. 3d ed. 3 vols. New York: Oxford University Press, 1996.

1. What Was Iron Doing at Life's Birth?

Alberts, B., D. Bray, J. Lewis, M. Raff, K. Roberts, and J. D. Watson. *Molecular Biology of the Cell*. 3d ed. New York: Garland Publishing, 1994.

Hamblin, W. K. *The Earth's Dynamic Systems: A Textbook in Physical Geology*. 5th ed. New York: Macmillian, 1989.

Pace, N. R. "A Molecular View of Microbial Diversity and the Biosphere." *Science* 276 (1997): 734–740.

Snow, T. P., K. R. Brownsberger. *Universe: Origins and Evolutions*. Belmont, Calif.: Wadsworth Publishing, 1997.

Whitman, W. B., D. C. Coleman, and W. J. Wiebe. "Prokaryotes: The

Unseen Majority." *Proceedings National Academy of Sciences USA* 95 (1998): 6,578–6,583.

Williams, R. J. P., and J. J. R. Frausto da Silva. *The Natural Selection of the Chemical Elements: The Environment and Life's Chemistry.* New York: Oxford University Press, 1996.

Woese, C. R. "Archaebacteria." *Scientific American* 244 (June 1981): 98–122.

———"The Universal Ancestor." *Proceedings National Academy of Sciences USA* 95 (1998): 6,854–6,859.

Woese, C. R., O. Kandler, and M. L. Wheelis. "Towards a Natural System of Organisms: Proposal for the Domains Achaea, Bacteria, and Eucarya." *Proceedings National Academy of Sciences USA* 87 (1990): 4,576–4,579.

2. Catastrophe

Adams, M.W.W., and E. I. Stiefel. "Biological Hydrogen Production: Not So Elementary." *Science* 282 (1998): 1,842–1,843.

Awramik, S. M. "Archaean and Proterozoic Stromatolites." In *Calcareous Algae and Stromatolites,* ed. R. Riding, 289–304. Berlin: Springer-Verlag, 1991.

Beinert, H., R. H. Holm, and E. Münck. "Iron-Sulfur Clusters: Nature's Modular, Multipurpose Structures." *Science* 272 (1997): 653–659.

Blakeslee, S. "Carl R. Woese: Bacteria's Solitary, Steadfast Champion." *New York Times,* October 15, 1996.

Broad, W. J. "Microbes Now Seen to Play Major Role in Shaping Earth's Crust." *New York Times,* October 15, 1996.

———"Comet's Gift: Hints of How Earth Came to Life." *New York Times,* April 1, 1997.

Browne, M. W. "Evidence Puts Date for Life's Origin Back Millions of Years." *New York Times,* November 7, 1996.

Cairns-Smith, A. G. *Genetic Takeover and the Mineral Origins of Life.* New York: Cambridge University Press, 1982.

Cano, R. J., and J. S. Colome. *Microbiology.* St. Paul, Minn.: West Publishing Company, 1986.

Corliss, J. B., J. A. Baross, and S. E. Hoffman. "An Hypothesis concerning the Relationship between Submarine Hot Springs and the Origin of

Life on Earth." *Oceanologica Acta. Proceedings 26th International Geological Congress* (1981): 59–69.

Crichton, R. P. *Inorganic Biochemistry of Iron Metabolism.* West Sussex, U.K.: Ellis Horwood Limited, 1991.

Delaney, J. R., D. S. Kelley, M. D. Lilley, D. A. Butterfield, J. A. Baross, W .S. D. Wilcock, R. W. Embley, and M. Summit. "The Quantum Event of Oceanic Crustal Accretion: Impacts of Diking at Mid-Ocean Ridges." *Science* 281 (1998): 222–230.

Ehrlich, H. *Geomicrobiology.* 3d ed. New York: Marcel Dekker, 1996.

Fassbinder, J. W. E., H. Stanjek, and H. Vali. "Occurrence of Magnetic Bacteria in Soil." *Nature* 343 (1990): 161–163.

Fredrickson, J. K., and T. C. Onstott. "Microbes Deep inside the Earth." *Scientific American* 275 (October 1996): 68–73.

Fyfe, W. S. "The Biosphere Is Going Deep." *Science* 273 (1996): 448.

Huber, C., and G. Wächtershäuser. "Peptides by Activation of Amino Acids with CO on (Ni,Fe)S Surfaces: Implications for the Origin of Life." *Science* 281 (1998): 670–672.

Kerr, R. A. "Life Goes to Extremes in the Deep Earth—and Elsewhere?" *Science* 276 (1997): 703–704.

Lipkin, R. "From Proteins to Protolife." *Science News* 146 (1994): 58–59.

Lovley, D. R. "Dissimilatory Fe(III) and Mn(IV) Reduction." *Microbiological Review* 55 (1991): 259–287.

———"Dissimilatory Metal Reduction." *Annual Review of Microbiology* 47 (1993): 263–290.

Margulis, L. *Symbiosis in Cell Evolution: Microbial Communities in the Archaean and Proterozoic Eons.* 2d ed. New York: W. H. Freeman, 1993.

McCollom, T. M., and E. L. Shock. "Geochemical Constraints on Chemolithoautotrophic Metabolism by Microorganisms in Seafloor Hydrothermal Systems." *Geochimica et Cosmochimica Acta* 61 (1997): 4,375–4,391.

Miller, S. L. "A Production of Amino Acids under Possible Primitive Earth Conditions." *Science* 117 (1953): 528–529.

Monastersky, R. "Signs of Ancient Life in Deep, Dark Rock." *Science News* 152 (1997): 181.

Morell, V. "Microbiology's Scarred Revolutionary." *Science* 276 (1997): 699–702.

Nash, J. M. "How Did Life Begin?" *Time,* October 11, 1993, 68–74.

Nealson, K. H., and C. R. Myers. "Microbial Reduction of Manganese and Iron: New Approaches to Carbon Cycling." *Applied and Environmental Microbiology* 58 (1992): 439–443.

Sawyer, K. "'Small-Game Hunting' on the Ocean Floor." *Washington Post*, August 28, 1998.

Simpson, S. "Life's First Scalding Steps." *Science News* 155 (1999): 24–26.

Vargas, M., K. Kashefi, E. L. Blunt-Harris, and D. R. Lovley. "Microbiological Evidence for Fe(III) Reduction on Early Earth." *Nature* 395 (1998): 65–67.

Wächtershäuser, G. "Life in a Ligand Sphere." *Proceedings National Academy of Sciences USA* 91 (1994): 4,283–4,287.

Wade, N. "Gunter Wächtershäuser: Amateur Shakes Up on Recipe for Life." *New York Times*, April 22, 1997.

Wiley, J. P. "Phenomena, Comment, and Notes." *Smithsonian* (June 1997): 22–24.

Williams, R. J. P. "Iron and the Origin of Life." *Nature* 343 (1990): 213–214.

3. Grabbing and Storing

Andrews, S. C., J. B. C. Findlay, J. R. Guest, P. M. Harrison, J. N. Keen, and J. M. A. Smith. "Physical, Chemical, and Immunological Properties of the Bacterioferritins of *Escherichia Coli, Pseudomonas Aeruginosa*, and *Azotobacter Vinelandii*." *Biochimica et Biophysica Acta* 1078, no. 1 (1991): 111–116.

Barton, L. L., and B. C. Hemming, eds. *Iron Chelation in Plants and Soil Microorganisms*. San Diego, Calif.: Academic Press, 1993.

Bauminger, E. R., and I. Nowik. "Magnetism in Plant and Mammalian Ferritin." *Hyperfine Interactions* 50 (1989): 489–497.

Bergeron, J. R., and G. M. Brittenham, eds. *The Development of Iron Chelators for Clinical Use*. Boca Raton, Fla.: CRC Press, 1994.

Bothwell, T. H. "Iron Overload in the Bantu." In *Iron Metabolism*, ed. F. Gross, 362–372. New York: Springer-Verlag, 1964.

Crichton, R. R. *Inorganic Biochemistry of Iron Metabolism*. New York: Ellis Horwood, 1991.

Crichton, R. R., and M. Charloteaux-Wauters. "Iron Transport and Storage." *European Journal of Biochemistry* 164 (1987): 485–506.

Dunford, H. B., D. Dolphin, K. N. Raymond, L. Sieker, eds. *The Biologi-*

cal Chemistry of Iron: A Look at the Metabolism of Iron and Its Subsequent Uses in Living Organisms: Proceedings of the NATO Advanced Study Institute, Held at Edmonton, Alberta, Canada, August 23–September 4, 1981. Dodrecht, Holland: D. Reidel, 1982.

Egan, T. J., and P. Aisen. "Metal-Transferrin Interactions." In *Iron Carriers and Iron Proteins,* ed. T. M. Loehr, 411–420. Physical Bioinorganic Chemistry Series, vol. 5. New York: VCH, 1989.

Emory, T. "Iron Metabolism in Humans and Plants." *American Scientist* 70 (1982): 626–632.

Ferguson, A. D., E. Hofmann, J. W. Coulton, K. Diederichs, and W. Welte. "Siderophore-Mediated Iron Transport: Crystal Structure of Fhua with Bound Lipopolysaccharide." *Science* 282 (1998): 2,215–2,220.

Harrison, P. M., P. J. Artymiuk, G. C. Ford, D. M. Lawson, J. M. A. Smith, A. Treffery, and J. L. White. "Ferritin: Function and Structural Design on an Iron-Storage Protein." In *Biomineralization: Chemical and Biochemical Perspectives,* ed. S. Mann and J. Webb, 257–294. New York: VCH, 1989.

Harrison, P. M., and T. H. Lilley. "Ferritin." In *Iron Carriers and Iron Proteins,* ed. T. M. Loehr, 125–238. Physical Bioinorganic Chemistry Series, vol. 5. New York: VCH, 1989.

Jacobs, I. S., and C. P. Bean. "Fine Particles, Thin Films, and Exchange Anisotropy (Effects of the Finite Dimensions and Interfaces on the Basic Properties of Ferromagnets)." In *Magnetism,* ed. G. T. Rado and H. Suhl, 3:271–349. New York: Academic Press, 1963.

Jiang, X., M. A. Payne, Z. Cao, S. B. Foster, J. B. Feix, S. M. C. Newton, and P. E. Klebba. "Ligand-Specific Opening of a Gated-Porin Channel in the Outer Membrane of Living Bacteria." *Science* 276 (1997): 1,261–1,264.

Jones, D. H. "Mössbauer Spectroscopy and the Physics of Superparamagnetism." *Hyperfine Interactions* 47 (1989): 289–297.

Lovley, D. R., M. J. Baedecker, D. J. Lonergan, I. M. Cozzarelli, E. J. P. Phillips, and D. I. Siegel. "Oxidation of Aromatic Contaminants Coupled to Microbial Iron Reduction." *Nature* 339 (1989): 1–3.

Lovley, D. R., and E. J. P. Phillips. "Novel Mode of Microbial Energy Metabolism: Organic Carbon Oxidation Coupled to Dissimilatory Reduction of Iron or Manganese." *Applied and Environmental Microbiology* 54 (1988): 1,472–1,480.

Lovley, D. R., E. J. P. Phillips, and D. J. Lonergan. "Hydrogen and Formate Oxidation Coupled to Dissimilatory Reduction of Iron or Manganese by *Alteromonas Putrefaciens." Applied and Environmental Microbiology* 55 (1989): 700–706.

Lovley, D. R., J. F. Stolz, G. L. Nord Jr., and E. J. P. Phillips. "Anaerobic Production of Magnetite by a Dissimilatory Iron-Reducing Microorganism." *Nature* 330 (1987): 252–254.

Matzanke, B. F., G. Müller-Matzanke, and K. N. Raymond. "Siderophore-Mediated Iron Transport." In *Iron Carriers and Iron Proteins*, ed. T. M. Loehr, 3–121. Physical Bioinorganic Chemistry Series, vol. 5. New York: VCH, 1989.

Mielczarek, E. V., P. W. Royt, and J. Toth-Allen. "Microbial Acquisition of Iron." *Comments on Molecular and Cellular Biophysics* 6 (1989): 1–30.

Neilands, J. B. "A Crystalline Organo-Iron Pigment from a Rust Fungus (*Ustilago Sphaerogena*)." *Journal of the American Chemical Society* 74 (1952): 4,846–4,847.

———"Microbial Metabolism of Iron." In *Iron in Biochemistry and Medicine*, ed. A. Jacobs and M. Worwood, 2:529–572. London: Academic Press, 1980.

———"Microbial Iron Compounds." *Annual Review of Biochemistry* 50 (1981): 715–731.

Nikaido, H., and M. H. Saier Jr. "Transport Proteins in Bacteria: Common Themes in Their Design." *Science* 258 (1992): 936–942.

Ponka, P., R. C. Woodworth, and H. M. Schulman, eds. *Iron Transport and Storage.* Boca Raton, Fla.: CRC Press, 1990.

Rutz, J. M., J. Liu, J. A. Lyons, J. Goranson, S. K. Armstrong, N. A. McIntosh, J. B. Feix, and P. E. Klebba. "Formation of a Gated Channel by a Ligand-Specific Transport Protein in the Bacterial Outer Membrane." *Science* 258 (1992): 471–475.

St. Pierre, T. G., and J. Webb. "Ferritin and Hemosiderin: Structural and Magnetic Studies of the Iron Core." In *Biomineralization: Chemical and Biochemical Perspectives*, ed. S. Mann, J. Webb, and R.J.P. Williams, 1–64. New York: VCH, 1989.

Telford, J. R., and K. N. Raymond. "Siderophores." In *Comprehensive Supermolecular Chemistry*, ed. J. L. Atwood, J. E. D. Davies, D. D. MacNichol, and F. Vögtle, 1:245–266. Oxford, U.K.: Elsevier Science Ltd., 1996.

Theil, Elizabeth C. "Ferritin: Structure, Gene Regulation, and Cellular Function in Animals, Plants, and Microorganisms." *Annual Review of Biochemistry* 56 (1987): 289–315.

Watt, G. D., R. B. Frankel, G. C. Papaefthymiou, K. Spartalian, and E. I. Stiefel. "Redox Properties and Mössbauer Spectroscopy of *Azotobacter Vinelandii* Bacterioferritin." *Biochemistry* 25 (1986): 4,330–4,336.

Webb, J., and T. G. St. Pierre. "The Use and Potential of Mössbauer Spectroscopy in Studies of Biological Mineralization." In *Mössbauer Spectroscopy Applied to Inorganic Chemistry*, ed. G. J. Long and F. Grandjean, 3:1–54. New York: Plenum Press, 1989.

Winkelmann, G., D. van der Helm, and J. B. Neilands, eds. *Iron Transport in Microbes, Plants, and Animals.* New York: VCH, 1987.

4. The Smallest Living Magnets

Bazylinski, D. A., R. B. Frankel, and H. W. Jannasch. "Anaerobic Magnetite Production by a Marine, Magnetotactic Bacterium." *Nature* 334 (August 1988): 518–519.

Blakemore, R. P. "Magnetotactic Bacteria." *Annual Review of Microbiology* 36 (1982): 217–238.

Blakemore, R. P., and R. B. Frankel. "Magnetic Navigation in Bacteria." *Scientific American* 245 (December 1981): 58–65.

Blakemore, R. P., R. B. Frankel, and A. J. Kalmijn. "South-Seeking Magnetotactic Bacteria in the Southern Hemisphere." *Nature* 286 (1980): 384–385.

Bronk, B. V., and J. W. Longworth. "Magnetotactic Bacteria." *Comments on Molecular and Cellular Biophysics* 1 (1982): 293–310.

Chang, S.-B. R., and J. L. Kirschvink. "Magnetofossils, the Magnetization of Sediments, and the Evolution of Magnetite Biomineralization." *Annual Review of Earth and Planetary Sciences* 17 (1989): 169–195.

Delong, E. F., R. B. Frankel, and D. A. Bazylinski. "Multiple Evolutionary Origins of Magnetotaxis in Bacteria." *Science* 259 (1993): 803–806.

Denham, C. R., R. P. Blakemore, and R. B. Frankel. "Bulk Magnetic Properties of Magnetotactic Bacteria." *IEEE Transactions on Magnetics* MAG-16 (September 1980): 1,006–1,007.

Farina, M., H. Lins de Barros, D. Motta, S. Esquivel, and J. Danon. "Ultrastructure of a Magnetotactic Microorganism." *Biology of the Cell* (193) 48 (1983): 85–88.

Farina, M., D. Motta, S. Esquivel, and H. Lins de Barros. "Magnetic Iron-Sulphur Crystals from a Magnetotactic Microorganism." *Nature* 343 (1990): 256–258.

Fassbinder, J. W. E., H. Stanjek, and H. Vali. "Occurrence of Magnetic Bacteria in Soil." *Letters to Nature* 343 (1990): 161–163.

Frankel, R. B. "Magnetic Guidance of Organisms." *Annual Review of Biophysics and Bioengineering* 13 (1984): 85–103.

Frankel, R. B., and R. P. Blakemore. "Navigational Compass in Magnetic Bacteria." *Journal of Magnetism and Magnetic Materials* 15–18, no. 3 (1980): 1,562–1,564.

———eds. *Iron Biominerals.* New York: Plenum Press, 1991.

Frankel, R. B., R. P. Blakemore, and R. S. Wolfe. "Magnetite in Freshwater Magnetotactic Bacteria." *Science* 203 (1979): 1,355–1,356.

Heywood, B. R., D. A. Bazylinski, A. Garratt-Reed, S. Mann, R. B. Frankel. "Controlled Biosynthesis of Greigite (Fe_3S_4) in Magnetotactic Bacteria." *Die Naturwissenschaften* 77 (1984): 536–538.

Kalmijn, A. J. "Biophysics of Geomagnetic Field Detection." *IEEE Transactions on Magnetics* MAG-17 (January 1981): 1,113–1,124.

Kirschvink, J. L. "South-Seeking Magnetic Bacteria." *Journal of Experimental Biology* 86 (1980): 345–347.

Kirschvink, J. L., and S.-B. R. Chang. "Ultrafine-Grained Magnetite in Deep-Sea Sediments: Possible Bacterial Magnetofossils." *Geology* 12 (1984): 559–562.

Liu, S. V., J. Zhou, C. Zhang, D. R. Cole, M. Gajdarziska-Josifovska, and T. J. Phelps. "Thermophilic Fe(III)–Reducing Bacteria from the Deep Subsurface: The Evolutionary Implications. *Science* 277 (1997): 1,106–1,109.

Lovley, D. R. "Organic Matter Mineralization with the Reduction of Ferric Iron: A Review." *Geomicrobiology Journal* 5 (1987): 375–399.

Lovley, D. R., J. F. Stolz, G. L. Nord Jr., and E. J. P. Phillips. "Anaerobic Production of Magnetite by a Dissimilatory Iron-Reducing Microorganism." *Nature* 330 (1987): 252–253.

Lowenstam, H. A. "Minerals Formed by Organisms." *Science* 211 (1981): 1,126–1,131.

Mann, S., N. H. C. Sparks, R. B. Frankel, D. A. Bazylinski, and H. W. Jannasch. "Biomineralization of Ferrimagnetic Greigite (Fe_3S_4) and Iron Pyrite (FeS_2) in a Magnetotactic Bacterium." *Nature* 343 (1990): 258–261.

Matsuda, T., J. Endo, N. Osakabe, and A. Tonomura. "Morphology and Structure of Biogenic Magnetite Particles." *Nature* 302 (1983): 411–412.

Moench, T. T. "Bilophococcus Magnetotacticus Gen. Nov. Sp. Nov., A

Motile, Magnetic Coccus." *Antonie van Leeuwenhoek (Journal of Microbiology and Serology)* 54 (1989): 483–496.

Moench, T. T., and W. A. Konetzka. "A Novel Method for the Isolation and Study of a Magnetotactic Bacterium." *Archives of Microbiology* 119 (1978): 203–212.

Monastersky, R. "The Flap over Magnetic Flips." *Science News* 143 (1993): 378–380.

Roden, E. E., and D. R. Lovley. "Dissimilatory Fe(III) Reduction by the Marine Microorganism *Desulfuromonas Acetoxidans.*" *Applied and Environmental Microbiology* 59 (1993): 734–742.

Rodgers, F. G., R. P. Blakemore, N. A. Blakemore, R. B. Frankel, D. A. Bazylinski, D. Maratea, and C. Rodgers. "Intercellular Structure in a Many-Celled Magnetotactic Prokaryote." *Archives of Microbiology* 154 (1990): 18–22.

Sarikaya, M., and I. A. Aksay, eds. *Biomimetics: Design and Processing of Materials.* Woodbury, N.Y.: AIP Press, 1995.

Sparks, N. H. C., S. Mann, D. A. Bazylinski, D. R. Lovley, H. W. Jannasch, and R. B. Frankel. "Structure and Morphology of Magnetite Anaerobically Produced by a Marine Magnetotactic Bacterium and a Dissimilatory Iron-Reducing Bacterium." *Earth and Planetary Science Letters* 98 (1990): 14–22.

Torres de Araujo, F. F., M. A. Pires, R. B. Frankel, and C.E.M. Bicudo. "Magnetite and Magnetotaxis in Algae." *Biophysical Journal* 50 (1986): 375–378.

5. Hemoglobin and Myoglobin

Adler, T. "Improving Humans' Blood with Crocodiles." *Science News* 147 (1995): 36.

Barnes, R.S.K., P. Calow, P.J.W. Olive, and D. W. Golding. *The Invertebrates: A New Synthesis.* 2d ed. Boston: Blackwell Science, 1993.

Bauman, R., H. Bartels, and C. Bauer. "Blood Oxygen Transport." In *Handbook of Physiology. The Respiratory System,* ed. L. E. Farhi and S. M. Tenney, 4: 147–172. Bethesda, Md.: American Physiological Society, 1987.

Benedek, G., and F.M.H. Villars. *Physics, with Illustrative Examples from Medicine and Biology.* Vol. 1, *Mechanics.* 2d ed. New York: Springer-Verlag, 2000.

Bowman, J. E., and E. Goldwasser. "Sickle Cell Fundamentals." A publication of the National Sickle Cell Disease Program, National Institutes of Health, 1975.

Broad, W. J. "Deep Underwater, the Breath of Life." *New York Times*, November 11, 1997.

Brody, J. E. "Taking Malaria Seriously." *International Herald Tribune*, October 5, 1998.

"Clinical Alert from the National Heart, Lung, and Blood Institute: The Stroke Prevention Trial in Sickle Cell Anemia." National Institutes of Health, September 18, 1997.

"Clinical Alert from the National Heart, Lung, and Blood Institute: Study of Hydroxyurea in Sickle Cell Anemia." National Institutes of Health, January 30, 1995.

Cooley's Anemia: Progress in Biology and Medicine—1995. Division of Blood Diseases and Resources, National Institutes of Health, 1995.

Dickerson, R. E., and I. Geis. *The Structure and Action of Proteins.* Menlo Park, Calif.: W. A. Benjamin, 1969.

Dietz, T. E. "More about Altitude Illness," http://www.gorge.net/hamg/AMS.html, March 6, 1996.

Eaton, W. H., and J. Hofrichter. "Hemoglobin S Gelation and Sickle Cell Disease." *Blood* 70 (1987): 1,245–1,266.

"FDA Approves Cancer Drug for Sickle Cell Anemia." *Washington Post*, March 5, 1998.

Ferrone, F. A. "The Polymerization of Sickle Hemoglobin in Solutions and Cells." *Experientia* 49 (1993): 110–117.

Kanwisher, J. W., and S. H. Ridgway. "The Physiological Ecology of Whales and Porpoises." *Scientific American* 248 (June 1983): 111–120.

Klocke, R. A., and A. R. Saltzman. "Gas Transport." In *Textbook of Pulmonary Disease,* ed. G. Baum and E. Wolinsky, 173–194. Boston: Little, Brown, 1989.

Klotz, I.M.G., L. Klippenstein, and W. A. Hendrickson. "Hemerythrin: Alternative Oxygen Carrier." *Science* 192 (1976): 335–344.

Komiyama, N. H., G. Miyazaki, J. Tame, and K. Nagai. "Transplanting a Unique Allosteric Effect from Crocodile into Human Haemoglobin." *Nature* 373 (1995): 244–246.

Kooyman, G. L., and P. J. Ponganis. "The Challenges of Diving to Depth." *American Scientist* 85 (1997): 530–539.

Lamy, J., J.-P. Truchot, and R. Gilles, eds. *Respiratory Pigments in Animals: Relation, Structure-Function.* New York: Springer-Verlag 1985.

Le Bouf, B. J. "Incredible Diving Machines." *Natural History* 98 (1989): 35–40.

Leary, W. E. "Transfusions Prevent Strokes in Children with Sickle Cell Disease, Study Finds." *New York Times,* September 19, 1997.

Lipsyte, R. "Competition and Drugs: Just Say Yes." *New York Times,* August 2, 1998.

Longman, J. "Lifesaving Drug Can Be Deadly When Misused." *New York Times,* August 9, 1998.

Macalpine, I., and R. Hunter. "Porphyria and King George III." *Scientific American* 221 (July 1969): 38–46.

Nathan, D. G. *Genes, Blood, and Courage: A Boy Called Immortal Sword.* Cambridge: Harvard University Press, 1995.

Nucci, M. L., and A. Abuchowski. "The Search for Blood Substitutes." *Scientific American* 278 (February 1998): 73–77.

Okie, S. "Genetics: A Youthful Trade-Off." *Washington Post,* October 14, 1996.

Perutz, M. F. "The Hemoglobin Molecule." *Scientific American* 211 (November 1964): 39–51.

———"Hemoglobin Structure and Respiratory Transport." *Scientific American* 239 (December 1978): 92–123.

Platt, O. S., D. J. Brambilla, W. F. Rosse, P. F. Milner, O. Castro, M. H. Steinberg, and P. P. Klug. "Mortality in Sickle Cell Disease: Life Expectancy and Risk Factors for Early Death." *New England Journal of Medicine* 330 (1994): 1,639–1,644.

Propper, R., and D. Nathan. "Clinical Removal of Iron." *Annual Review of Medicine* 33 (1982): 509–519.

Spiro, T. G., and W. M. Stigliani. *Chemistry of the Environment.* Upper Saddle River, N.J.: Prentice Hall, 1996.

Stamatoyannopoulos, G. "The Molecular Basis of Hemoglobin Disease." *Annual Review of Genetics* 6 (1972): 47–69.

Stamatoyannopoulos, G., A. J. Bellingham, C. Lenfant, and C. A. Finch. "Abnormal Hemoglobins with High and Low Oxygen Affinity." *Annual Review of Medicine* 2 (1971): 221–234.

Stryer, Lubert. *Molecular Design of Life.* New York: W. H. Freeman, 1989.

———*Biochemistry.* 3d ed. New York: W. H. Freeman, 1995.

Sunshine, H. R., J. Hofrichter, and W. A. Eaton. Requirements for Therapeutic Inhibition of Sickle Haemoglobin Gelation." *Nature* 275 (1978): 238–240.

Vullo, R., B. Modell, and E. Georganda. *What Is Cooley's Anemia?* 2d ed. Flushing, N.Y.: Cooley's Anemia Foundation, 1995.

Weiss, R. "Sickle Cell Treatment Prevents Strokes in Children." *Washington Post,* September 19, 1997.

————"Transplants Appear to Cure Sixteen with Sickle Cell Disease." *Washington Post,* August 8, 1996.

Weissbluth, M. "The Physics of Hemoglobin." *Structure and Bonding.* New York: Springer-Verlag, 1977.

6. Migrating Animals

Abbott, S. *The Bookmaker's Daughter.* New York: Houghton Mifflin, 1991.

Able, K. P. "Magnetic Orientation and Magnetoreception in Birds." *Progress in Neurobiology* 42 (1994): 449–473.

Able, K. P., and M. A. Able. "Interactions in the Flexible Orientation System of a Migratory Bird." *Nature* 375 (1995): 230–232.

————"Manipulations of Polarized Skylight Calibrate Magnetic Orientation in a Migratory Bird." *Journal of Comparative Physiology* 177 (1995): 351–356.

Beason, R. C., and P. Semm. "Detection of and Receptors for Magnetic Fields in Birds." *Biological Effects of Electric and Magnetic Fields* 1 (1994): 241–260.

Brown, F. A., Jr., F. H. Barnwell, and H. Marguerite Webb. "Adaptation of the Magnetoreceptive Mechanism of Mud-Snails to Geomagnetic Strength." *Biological Bulletin* 127 (1964): 221–231.

Brown, F. A., Jr., and Y. H. Park. "Duration of an After-Effect in Planarians following a Reversed Horizontal Magnetic Vector." *Biological Bulletin* 128 (1965): 347–355.

Chau-anusorn, W., T. G. St. Pierre, G. Black, J. Webb, D. J. Macey, and D. Parry. "Mössbauer Spectroscopic Study of Iron Oxide Deposits in Liver Tissue from the Marine Mammal, Dugong dugong." *Hyperfine Interactions* 91 (1994): 899–904.

Goff, M., M. Salmon, and K. J. Lohmann. "Hatching Sea Turtles Use Surface Waves to Establish a Magnetic Compass Direction." *Animal Behavior* 55 (1998): 69–77.

Gould, J. L. "The Case for Magnetic Sensitivity in Birds and Bees (Such As It Is)." *American Scientist* 68 (1980): 256–267.

————"The Map Sense of Pigeons." *Nature* 296 (1982): 205–211.

Gould, J. L., and K. P. Able. "Human Homing: An Elusive Phenomenon." *Science* 212 (1981): 1,061–1,063.

Gould, J. L., J. L. Kirschvink, and K. S. Deffeyes. "Bees Have Magnetic Remanence." *Science* 201 (1978): 1,026–1,028.

Gould, J. L., J. L. Kirschvink, K. S. Deffeyes, and M. L. Brines. "Orientation of Demagnetized Bees." *Journal of Experimental Biology* 86 (1980): 1–8.

Kirschvink, J. L. "The Horizontal Magnetic Dance of the Honeybee Is Compatible with a Single-Domain Ferromagnetic Magnetoreceptor." *BioSystems* 14 (1981): 193–203.

———"Magnetite Biomineralization and Geomagnetic Sensitivity in Higher Animals: An Update and Recommendations for Future Study." *Bioelectromagnetics* 10 (1989): 239–260.

Kirschvink, J. L., and J. L. Gould. "Biogenic Magnetite as a Basis for Magnetic Field Detection in Animals." *BioSystems* 13 (1981): 181–201.

Kirschvink, J. L., and H. A. Lowenstam. "Mineralization and Magnetization of Chiton Teeth: Paleomagnetic, Sedimentologic, and Biologic Implications of Organic Magnetite." *Earth and Planetary Science Letters* 44 (1979): 193–204.

Light, P., M. Salmon, and K. J. Lohmann. "Geomagnetic Orientation of Loggerhead Sea Turtles: Evidence for an Inclination Compass." *Journal of Experimental Biology* 182 (1993): 1–10.

Lohmann, K. J., and C.M.F. Lohmann. "Detection of Magnetic Field Intensity by Sea Turtles." *Nature* 380 (1996): 59–61.

———"Sea Turtle Navigation and the Detection of Geomagnetic Field Features." *Journal of Navigation* 50 (1998): 10–22.

Lohmann, K. J., N. D. Perntcheff, G. A. Nevitt, G. D. Stetten, R. K. Zimmer-Faust, H. E. Jarrard, and L. C. Boles. "Magnetic Orientation of Spiney Lobsters in the Ocean: Experiments with Undersea Coil Systems. *Journal of Experimental Biology* 198 (1995): 2,041–2,048.

Lohmann, K. J., B. E. Witherington, C.M.F. Lohmann, and M. Salmon. "Orientation, Navigation, and Natal Beach Homing in Sea Turtles." In *The Biology of Sea Turtles,* ed. P. L. Lutz and J. A. Musick, 107–135. Boca Raton, Fla.: CRC Press, 1996.

Mann, S., N. H. C. Sparks, M. M. Walker, and J. L. Kirschvink. "Ultrastructure, Morphology, and Organization of Biogenic Magnetite from Sockeye Salmon, *Oncorhynchus Nerka:* Implications for Magnetoreception." *Journal of Experimental Biology* 140 (1988): 35–49.

Martin, H., and M. Lindauer. "The Effect of the Earth's Magnetic Field on Gravity Orientation in the Honey Bee (*Apis Mellifira*)." *Journal of Comparative Physiology* 122 (1977): 145–187.

Meldrum, F. C., B. R. Heywood, D.P.E. Dickson, and S. Mann. "Iron Biomineralization in the Poriferan *Ircinia Oros*." *Journal of the Marine Biology Association of the U.K.* 75 (1995): 993–996.

Metcalfe, J. D., B. H. Holfor, and G. P. Arnold. "Orientation of Plaice (*Pleuronectes Platessa*) in the Open Sea: Evidence for the Use of External Directional Clues." *Marine Biology* 117 (1993): 559–566.

Presti, D., and J. D. Pettigrew. "Ferromagnetic Coupling to Muscle Receptors as a Basis for Geomagnetic Field Sensitivity in Animals." *Nature* 285 (1980): 99–101.

Quinn, T. P. "Evidence for Celestial and Magnetic Compass Orientation in Lake Migrating Sockeye Salmon Fry." *Journal of Comparative Physiology A: Sensory, Neural, and Behavioral Physiology* 137 (1980): 243–248.

Schmidt-Koenig, K. "Bird Navigation: Has Olfactory Orientation Solved the Problem?" *The Quarterly Review of Biology* 62 (1987): 31–47.

Schmidt-Koenig, K., and W. T. Keeton, eds. *Animal Migration, Navigation, and Homing.* New York: Springer-Verlag, 1978.

Schreiber, B., and O. Rossi. "Correlation between Race Arrivals of Homing Pigeons and Solar Activity." *Bollettino di Zoologia* 43 (1976): 317–320.

————"Correlation between Magnetic Storms Due to Solar Spots and Pigeon Homing Performances." *Scientific Review* 14 (1978): 961–963.

Towe, K. M., and H. A. Lowenstam. "Ultrastructure and Development of Iron Mineralization in the Radular Teeth of *Cryptochiton Stelleri* (Mollusca)." *Journal of Ultrastructure Research* 17 (1967): 1–13.

Walcott, C., J. L. Gould, and J. L. Kirschvink. "Pigeons Have Magnets." *Science* 205 (1979): 1,027–1,029.

Walcott, C., and R. P. Green. "Orientation of Homing Pigeons Altered by a Change in the Direction of an Applied Magnetic Field." *Science* 184 (1974): 180–182.

Walker, M. M., and M. E. Bitterman. "Magnetic Sensitivity in Honeybees: Threshold Estimation and Test of a Mechanism." In *Effects of Atmospheric and Geophysical Variables in Biology and Medicine,* ed. H. Lieth, 53–56. The Hague, the Netherlands: SPB Academic Publishing, 1991.

Walker, M. M., C. E. Diebel, C. V. Haugh, P. M. Pankherst, J. C. Montgomery, and C. R. Green. "Structure and Function of the Vertebrate Magnetic Sense." *Nature* 390 (1997): 371–376.

Walker, M. M., T. P. Quinn, J. L. Kirschvink, and C. Groot. "Production of Single-Domain Magnetite throughout the Life by Sockeye Salmon, *Oncorhynchus Nerka*." *Journal of Experimental Biology* 140 (1988): 51–64.

Wiltschko, R., and W. Wiltschko. *Magnetic Orientation in Animals.* Zoophysiology, vol. 33. New York: Springer-Verlag, 1995.

Wiltschko, W. "The Influence of Magnetic Total Intensity and Inclination on Directions Preferred by Migrating European Robins." In *Symposium NASA SP–262: Animal Orientation and Navigation,* 569–578. Washington, D.C.: U.S. Government Printing Office, 1972.

Wiltschko, W., U. Munro, H. Ford, and R. Wiltschko. "Red Light Disrupts Magnetic Orientation of Migratory Birds." *Nature* 364 (1993): 525–527.

————"Magnetic Inclination Compass: A Basis for the Migratory Orientation of Birds in the Northern and Southern Hemisphere." *Experientia* 49 (1993): 167–170.

Wiltschko, W., D. Nohr, E. Fuller, and R. Wiltschko. "Pigeon Homing: The Use of Magnetic Information in Position Finding." In *Biophysical Effects of Steady Magnetic Fields,* ed. G. Maret, N. Boccara, and J. Kiepenheuer, 154–162. New York: Springer, 1985.

Wiltschko, W., and R. Wiltschko. "Bird Orientation under Different Sky Sectors." *Zeitschrift der Tierpsychologie* 35 (1974): 536–542.

————"Migratory Orientation of European Robins Is Affected by the Wavelength of Light as Well as by Magnetic Pulse." *Journal of Comparative Physiology A: Sensory, Neural, and Behavioral Physiology* 177 (1995): 363–369.

————"Magnetic Orientation in Birds." *Journal of Experimental Biology* 199 (1996): 29–38.

Yeagley, H. L. "A Preliminary Study of a Physical Basis of Bird Navigation." *Journal of Applied Physics* 18 (1947): 1,035–1,063.

————"A Preliminary Study of a Physical Basis of Bird Navigation. Part II." *Journal of Applied Physics* 22 (1951): 746–760.

Yorke. E. D. "Sensitivity of Pigeons to Small Magnetic Field Variations." *Journal of Theoretical Biology* 89 (1981): 533–537.

Zoeger, J., J. R. Dunn, and M. Fuller. "Magnetic Material in the Head of the Common Pacific Dolphin." *Science* 213 (1981): 892–894.

7. Iron and the Planet's Ecosystem

Appleby, C. A., E. S. Dennis, and W. J. Peacock. "A Primaeval Origin for Plant and Animal Haemoglobins?" *Australian Systematic Botany* 3 (1990): 81–89.

Barton, L., and B. Hemming, eds. *Iron Chelation in Plants and Soil Microorganisms.* San Diego, Calif.: Academic Press, 1993.

Behrenfeld, M. J., A. J. Bale, Z. S. Kobler, J. Aiken, and P. G. Falkowski. "Confirmation of Iron Limitation of Phytoplankton Photosynthesis in the Equatorial Pacific Ocean." *Nature* 383 (1996): 508–511.

Booth, William. "Ironing Out 'Greenhouse Effect': Fertilizing Oceans Is Proposed to Spur Algae." *Washington Post,* May 20, 1990.

Brill, W. J. "Biological Nitrogen Fixation." *Scientific American* (March 1997): 68–82.

Chisholm, S. "The Iron Hypothesis: Basic Research Meets Environmental Policy." *Reviews of Geophysics, Suppl.* (July 1995): 1,277–1,286.

Chisholm, S., and F. Morel, eds. "What Controls Phytoplankton Production in Nutrient-Rich Areas of the Open Sea?" *Limnology and Oceanography* 36, special issue (December 1991).

Coale, K. H., K. S. Johnson, S. E. Fitzwater, R. M. Gordon, S. Tanner, F. P. Chavez, L. Ferioli, C. Sakamoto, P. Rogers, F. Miller, P. Stenberg, P. Nightingale, D. Cooper, W. P. Cochlan, M. R. Landry, J. Constantinou, G. Rollwagen, A. Trasvina, and R. Kudela. "A Massive Phytoplankton Bloom Induced by an Ecosystem-Scale Iron Fertilization Experiment in the Equatorial Pacific Ocean." *Nature* 383 (1996): 495–501.

de Baar, H. J. W., J. T. M. de Jong, D. C. E. Bakker, B. M. Löscher, C. Veth, U. Bathmann, and V. Smetacek. "Importance of Iron for Plankton Blooms and Carbon Dioxide Drawdown in the Southern Ocean." *Nature* 373 (1995): 412–415.

Dey, P. M., and J. B. Haborne, eds. *Plant Biochemistry.* San Diego, Calif.: Academic Press, 1997.

Environmental News Network Staff. "Iron Plays Key Role in Ocean CO_2 Absorption." CNN Interactive, June 15, 1998, http://cnn.com/TECH/science/9806/15/co2.yoto/index.html.

Fischer, R., and S. Long. "Rhizobium-Plant Signal Exchange." *Nature* 357 (1992): 655–660.

Frost, B. W. "Phytoplankton Bloom." *Nature* 383 (1996): 475–476.

Hutchins, D. A., and K. W. Bruland. "Iron-Limited Diatom Growth and

Si:N Uptake Ratios in a Coastal Upwelling Regime." *Nature* 393 (1998): 561–564.

Hutchins, D. A., G. R. DiTullio, Y. Zhang, and K. W. Bruland. "An Iron Limitation Mosaic in the California Upwelling Regime." *Limnology and Oceanography* 43 (1998): 1,037–1,054.

Kerr, R. A. "Iron Fertilization: A Tonic, but No Cure for the Greenhouse." *Science* 263 (1994): 1,089–1,090.

Kirchman, D. L. "Microbial Breathing Lessons." *Nature* 385 (1997): 121–122.

Long, S. R., and B. J. Staskawicz. "Prokaryotic Plant Parasites." *Cell* 73 (1993): 921–935.

Martin, J. H. "Glacial Interglacial CO_2 Change: The Iron Hypothesis." *Paleoceanography* 5 (1990): 1–13.

Martin, J. H., K. H. Coale, K. S. Johnson, S. E. Fitzwater, R. M. Gordon, S. J. Tanner, C. N. Hunter, V. A. Elrod, J. L. Nowicki, T. L. Coley, R. T. Barber, S. Lindley, A. J. Watson, K. Van Scoy, C. S. Law, M. I. Liddicoat, R. Ling, T. Stanton, J. Stockel, C. Collins, A. Anderson, R. Bidigare, M. Ondrusek, M. Latasa, F. J. Millero, K. Lee, W. Yao, J. Z. Zhang, G. Friedrich, C. Sakamoto, F.Chavez, K. Buck, Z. Kolber, R. Greene, P. Falkowski, S. W. Chisholm, F. Hoge, R. Swift, J. Yungel, S. Turner, P. Nightingale, A. Hatton, P. Liss, and N. W. Tindale. "Testing the Iron Hypothesis in Ecosystems of the Equatorial Pacific Ocean." *Nature* 371 (1994): 123–129.

Monastersky, R. "Iron versus the Greenhouse: Oceanographers Cautiously Explore a Global Warming Therapy." *Science News* 148 (1995): 220–221.

Mortlock, R. A., C. D. Charles, P. N. Froelich, M. A. Zibello, J. Saltzman, J. D. Hays, and L. H. Burckle. "Evidence for Lower Productivity in the Antarctic Ocean during the Last Glaciation." *Nature* 351 (1991): 220–222.

Sigel, A., and H. Sigel, eds. *Metal Ions in Biological Systems.* New York: Marcel Dekker 1998.

Smil, V. "Global Population and the Nitrogen Cycle." *Scientific American* 277 (July 1997): 76–81.

Stevens, W. K. "Too Much of a Good Thing Turns Nitrogen into a Threefold Menace." *New York Times,* December 10, 1996.

Suplee, C. "Iron Supplement for Cooler World." *Washington Post,* October 14, 1996.

Szalai, V. A., and G. W. Brudvig. "How Plants Produce Dioxygen." *American Scientist* 86 (1998): 543–551.

Tortell, P. D., M. T. Maldonado, and N. M. Price. "The Role of Heterotrophic Bacteria in Iron-Limited Ocean Ecosystems." *Nature* 383 (1996): 330–332.

Van Scoy, K., and K. Coale. "Pumping Iron in the Pacific." *New Scientist* 144 (1994): 32–35.

Wittenburg, J. B., and B. A. Wittenburg. "Mechanisms of Cytoplasmic Hemoglobin and Myoglobin Function." *Annual Review of Biophysical Chemistry* 19 (1990): 217–241.

8. Feeding the World's Poor

Becroft, D.M.O., M. R. Dix, and K. Farmer. "Intramuscular Iron-Dextran and Susceptibility of Neonates to Bacterial Infections." *Archives of Diseases in Childhood* 52 (1977): 778–781.

Brock, J. H., J. W. Halliday, M. J. Pippard, and L. W. Powell, eds. *Iron Metabolism in Health and Disease.* Philadelphia: W. B. Saunders, 1994.

Brody, J. E. "In Vitamin Mania, Millions Take a Gamble on Health." *New York Times,* October 26, 1997.

Brown, J. L., and E. Pollitt. "Malnutrition, Poverty, and Intellectual Development." *Scientific American* 274 (February 1996): 38–43.

"Champion of Children." *Johns Hopkins Children's Center News* 9 (fall 1985): 1, 4.

Emery, T. "Iron Metabolism in Humans and Plants." *American Scientist* 70 (1982): 626–632.

———*Iron and Your Health: Facts and Fallacies.* Boca Raton, Fla.: CRC Press, 1991.

Fairweather-Tait, S. "Iron." *International Journal for Vitamin and Nutrition Research* 63 (1993): 296–301.

"Frank Aram Oski, 64, Noted Pediatrician." *Philadelphia Inquirer,* December 9, 1996.

Gordeuk, V. R., G. M. Brittenham, and P. E. Thuma. "Iron Chelation as a Chemotherapeutic Strategy in the Treatment of Malaria." In *The Development of Iron Chelators for Clinical Use,* ed. R. J. Bergeron and G. M. Brittenham, 111–130. Boca Raton, Fla.: CRC Press, 1994.

Gunshin, H., B. Mackenzie, U. V. Berger, Y. Gunshin, M. F. Romero, W. F. Boron, S. Nussberger, J. L. Gollan, and M. A. Hediger. "Cloning and Characterization of Mammalian Proton-Coupled Metal-Ion Transporter." *Nature* 388 (1997): 482–487.

Hercberg, S., and P. Galan. "Biochemical Effects of Iron Deprivation." *Acta Paediatrica Scandinavica Suppl.* 361 (1989): 63–70.

Honig, A. S., and F. A. Oski. "Developmental Scores of Iron Deficient Infants and the Effects of Therapy." *Infant Behavior and Development* 1 (1978): 168–176.

———"Solemnity: A Clinical Risk Index of Iron Deficient Infants." *Early Child Development and Care* 16 (1984): 69–84.

International Nutritional Anemia Consultative Group, ILSI Human Nutrition Institute. "Iron EDTA for Food Fortification." Washington, D.C., 1997.

Jurado, R. L. "Iron, Infections, and Anemia of Inflammation." *Clinical Infectious Diseases* 25 (1997): 888–895.

Kiester, E., Jr. "A Little Fever Is Good for You." *Science* 84 (1984): 168–173.

Lewin, D. I. "Evolutions: Free-Radical Protection Systems." *The Journal of NIH Research* 9 (1997): 72–67.

Looker, A. C., P. R. Dallman, M. D. Carroll, E. W. Gunter, and C. L. Johnson. "Prevalence of Iron Deficiency in the United States." *JAMA* 277 (1997): 973–976.

Lozoff, B., E. Jimenez, and A. W. Wolf. "Long-Term Developmental Outcome of Infants with Iron Deficiency." *New England Journal of Medicine* 35 (1991): 687–694.

Mabeza, G. F., G. Biemba, and V. R. Gordeuk. "Clinical Studies of Iron Chelators in Malaria." *Acta Haematologica* 95 (1996): 78–86.

Mandishona, E., A. P. MacPhail, V. R. Gordeuk, M.–A. Kedda, A. C. Paterson, T. A. Rouault, and M. C. Kew. "Dietary Iron Overload as a Risk Factor for Hepatocellular Carcinoma in Black Africans." *Hepatology* 27 (1998): 1,563–1,566.

Morris, C. J., J. R. Earl, C. W. Trenam, and D. R. Blake. "Reactive Oxygen Species and Iron—A Dangerous Partnership in Inflammation." *International Journal of Biochemistry and Cell Biology* 27 (1995): 109–122.

Moyo, V. M., I. T. Gangaidzo, V. R. Gordeuk, C. F. Kiire, and A. P. MacPhail. "Tuberculosis and Iron Overload in Africa: A Review." *Central African Journal of Medicine* 43 (1997): 334–339.

Murray, M. J., A. B. Murray, M. B. Murray, C. J. Murray. "The Adverse

Effect of Iron Repletion on the Course of Certain Infections." *British Medical Journal* 2 (1978): 1,113–1,115.

Newman, J. "How Breast Milk Protects Newborns." *Scientific American* 273 (December 1995): 76–79.

Oldenburg, D. "Iron: Too Much of a Good Thing?" *Washington Post,* July 23, 1997.

Oski, F. A., A. S. Honig, B. Helu, and P. Howanitz. "Effect of Iron Therapy on Behavior Performance in Nonanemic, Iron-Deficient Infants." *Pediatrics* 71 (1983): 877–880.

Pollitt, E. "Iron Deficiency and Cognitive Function." *Annual Reviews of Nutrition* 13 (1993): 521–527.

Pollitt, E., J. Haas, and D. A. Levitsky. "Iron Deficiency and Educational Achievement in Thailand." *American Journal of Clinical Nutrition* 50 (1989): 687–689.

Roncagliolo, M., M. Garrido, T. Walter, P. Peirano, and B. Lozoff. "Evidence of Altered Central Nervous System Development in Infants with Iron Deficiency Anemia at 6 Months: Delayed Maturation of Auditory Brainstem Responses." *American Journal of Clinical Nutrition* 68 (1998): 683–690.

Scrimshaw, N. S. "Iron Deficiency." *Scientific American* 265 (October 1991): 46–52.

Viteri, F. E. "Prevention of Iron Deficiency." In *Prevention of Micronutrient Deficiencies: Tools for Policy Makers and Public Health Workers,* ed. C. P. Howson, E. T. Kennedy, and A. Horwitz, 45–102. Washington, D.C.: National Academy Press, 1998.

Weinberg, E. D. "The Development of Awareness of Iron-Withholding Defense." *Perspectives in Biology and Medicine* 36 (1993): 215–221.

INDEX

Page numbers for illustrations appear in italics.

Able, Kenneth, 109, 110, 115, 116, 120
Able, Mary A., 120
aerobic power, 86
Allen, Paul, 10, 15, 32, 34, 85, 110, 139, 160
alpha chains, 75–76, 78–79, 89, 90, 95, 99, 101
Alvin (deep-sea submersible), 11–13, *18*, 20
amino acids, essential, 139, 140, *141*, 146, 148
ammonia, 10, 139, 147, 148
Anabena, 146
anaerobic: conditions, 6, 13, 15, 16, 19, 23, 28; environments, *2*, 24, 72; organisms, 19–21, 28, 146
anemia, 31, 44, 46, 73, 88, 90, 149–158; pernicious, 39, 77; and sickle cell disease, 95, 97; thalassemia, 99–101
anemone, 13
angular dip, of compass, 122. *See also* inclination
animals, 17, 19, *22*, 24, *26*, 72, 126, 134, 147, 155; hemoglobin in, 91, 92; magnetic orientation and, 105–109, 110, 120–123; magnetic sensory organ in, 127, 128
Aquaspirillum magnetotacticum, 53, 54, 65

Araujo, Flavio Torres de, 62
archaea, discovery of, 17–20, *22*, 23–25, 49, 54, 66, 145
Arctic tern, 105
atoms, 25, 139, 146–148. *See also specific elements, e.g., iron*

Bacillus infernus, 21
bacteria: evolution of, 17–20, *22*, 24, 25, *26*, 28; iron store balance in, 31, 33–36, 38–39, 46; cell, *35, 46,* 53; and magnetism, 49–70, 106, 109–110; magnetite-ejecting, *64,* 65; and nitrogen fixation, 131, 138–143, 145; and producing human hemoglobin, 103; in rock, 20–21
bacteroids, 143, *144,* 145
Ballard, Robert D., 13
banded iron formation, 27, 66
Bantu, iron overload in, 37
Barber, Richard, 138
Baross, John, 15
Bean, Charles A., 38
Beason, Robert C., 119
Bernal, J. D. "Sage," 10, 74
beta chains, 55, 75–76, 78–79, 89–90, 95, 97, 99, 101
beta thalassemia. *See* Cooley's anemia
bioengineering, 88, 98, 140, 153
biological scale, *37*

biomineralization, of iron compounds, 62, *64*, 65, 66, 126–128
birds, *26*, 44, 93, 105–121, 127–129, 131
Blakemore, Nancy, 52, 60
Blakemore, Richard P., 49–55, 56, 57, 59, 60, 63, 65
blood diseases,90–91, 95. *See also specific diseases*
blood doping, 87, 88
blue-green algae. *See* cyanobacteria
bobolink, American, 119
brackish water, 63, 65
breast milk, 151
Bresnick, John, 50

Cairns-Smith, A. Graham, 15
Campylobacter jejuni, 31
Canale-Parola, Ercole, 53, 55
capillaries: for embryonic and fetal hemoglobin, 88, *89, 90;* and hemoglobin transport, 72, 73, 78, 79, 82, 83, 84, 85; in high altitude, 86; in sickle cell disease, 97
carbon dioxide: in atmosphere, 25; in blood, 73, 84; and greenhouse effect, 131–135, 137–139; in marine mammals, 92; and respiration, 29
carbon monoxide, 25, 93
Caretta caretta. See loggerhead sea turtle
celestial cues, for migration, 106, 109, 121
cells: iron in, 31; magnetic bacteria, 51, 53, 59, 64, 65; muscle, 85, 128; and oxygen, 28, 29, 86, 93; plant, 133, 142–144, *144,* 146; magnetic receptor, 127; reactions in, 7, 9, 20; siderophores in, 32, 34, *35, 36;* transferrin in, 41, 42, 43. *See also* red blood cells
cereal (grains), 140, 148, 153
Chelonia mydas. See green turtle
chemistry: early biochemistry, 7, 9,

24; of hot springs, 11; oxygen-based, 28, 29; of sea vents, 12, 15, 21. *See also* iron
childproof packaging, of pills, 150
chinook salmon, 124–125
Chisholm, Sallie W., 135, 137
chiton, 51, 126, 127. *See also* biomineralization
chlorophyll, 19, 25, 77, 131, 132, 133, 137
chloroplasts, 133
Cryptochiton stelleri. See chiton
clay, 15
Coale, Kenneth, 25, 137, 138
cobalt, 55, 69, 77
Columbia livia. See rock dove
compasses: in atomic particles, 66; in bacteria, 51, 55–60; in birds, 106, 109–115, *112,* 118–121; and Earth's magnetic field, 61, 62, *112;* in other animals, 125, 126; in sea turtles, 118–123
Cooley's anemia, 99–101. *See also* thalassemia
cooperativity, 79–81
copper, 28, 91
Corliss, John, 11–15, 17
crocodile, 75, 91, 92
cyanide, 93
cyanobacteria, 21, 146
cytochrome *c,* 32
cytochrome oxidase, 28, 93
cytoplasm, 35, 36

deferoxamine, 100
Delphinus delphis. See Pacific dolphin
Desferal, 45, 100
diatoms, 138
dietary supplements (iron), 42, 46, 149, 150, 151, 152, 153, 154, 156. *See also* vitamins
dioxyribonucleic acid. *See* DNA
2,3-diphosphoglycerate (DPG), 87
diphtheria, 31, 45

DNA, 10, 20, *26*, 29, 42, 44, 96
Dolichonyx oryzivorus. See bobolink, American
domain. *See* magnetic domain
Donnelly, Jack, 12
dugong, 126
Dymond, Jack R., 12

E. coli, 33, 34, 35, 38, 154
Earth: core, 6, 15, 103, 107, 118, 128, 159; crust, 16; formation of, 5–11, 21, 25, 28–29. *See also* greenhouse effect; magnetic field lines
echolocation, 117
Edmond, John M., 13
EDTA, 33, 154
electron microscopy, 52–55, *58, 64,* 127
electron spin: and hemoglobin, 79, 80; and magnetism, *8,* 38, 57, 58, 66–70, *65, 66, 67*
electrons, 17; as energy source, 24, 32, 133; and free radicals, 28, 43; of iron, 6–9; and magnetic exchange interaction, *69;* and magnetism, 38, 57–58, 66, *67;* and oxygen docking, 79, *80. See also* electron microscopy
endocytosis, *43*
enterobactin, 33, *34*
epsilon chains, 89, *90*
Erithacus rubecula. See European robin
erythrocythemia, 88
erythropoietin (EPO), 88
Escherichia coli. See E. coli
essential amino acids, 140, 146, 148
ethylenediaminetetraacetic acid. *See* EDTA
eucaryote, 17, 20, *22*
Europa, 21
European blackcap, 119, 120
European garden warbler, 118, 119

European robin, 106, 107, 109, 111, 114, 115, 120
evolution of life, *22, 26, 27*
exchange interaction. *See* magnetic exchange interaction

Farrant, John, 39
ferric iron (F+++), 6, *8,* 9, 23–28, 33–34, 38, 65, 73, 79–80, 140
ferric oxide, 27, 33, 73
ferrihemoglobin, 73
ferritin, 31, *37,* 37–40, *41,* 44, 133; and hemochromatosis, 45–46, 73, 126, 133; and hemoglobin disorders, 157, 158; and magnetic domains, 38; and transferrin, 42–44
ferrous iron (F++), 6, *8,* 9, 24, 27–28, 38, 73, 79–80, 140, 154–155, 157
fertilizer, 140, 147, 148
Ficedula hypoleuca. See pied flycatcher
foods, iron content of, 153
fool's gold. *See* pyrite
fossil fuels, 132, 148
fossils, 20, 21, 59
Frankel, Richard, 55, 56, 57, 59, 60, 62, 63
free radicals, 28, 43
Fromme, Hans Gerhard, 107, 108

Galapagos Islands, 11, 137
gamma chains, 89, *90,* 99
gamma rays, 55, 56
genetic blood diseases, 45, 95, 99. *See also specific diseases*
Geobacter metallireducens, 64, 65
global warming, 131, 132, 134, 135, *136,* 138, 139
Godzilla tower, 16–18, *18*
goethite, 126
goldcrest, 119
Gould, James, 116
grains. *See* cereal
Granick, Sam, 40

Green, Robert, 115
green turtle, 122
greenhouse effect, 131, 132, 134, 137, 139
Griffin, Donald, 117
gyre. *See* North Atlantic gyre

Haber-Bosch process, 148
Harrison, Pauline, 39, 40
hematoporphyrin, 95
heme, 75, *77,* 77–80, *80,* 91, 94, 145
hemerythrin, 91
hemochromatosis, 31, 44, 45, 46, 159
hemoglobin, 13, 71–103, *76, 82, 90,* 157, 158; adult, 88, 90, 98; artificial, 101–103; embryonic, *90;* fetal, 88, *89, 90,* 91, 97, 98, 99, 101; and iron deficiency, 149, 151, 153, 155; maternal, *89;* plant, 140, 143, *144,* 145
heterocysts, 146
high altitude, and hemoglobin, 85–88
high nutrient–low chlorophyll (HNLC) areas, 131, 132, 133, 136
HNLC. *See* high nutrient–low chlorophyll
Hodgkin, Dorothy Crowfoot, 40, 74
homing, 105, 110, 113–115, *116,* 127. *See also* pigeon
honeybee, 123, 125
honeyeater, yellow-faced, 119
Honig, Alice S., 155–156
hookworm, 152, 157, 158
horse spleen, 40
horseshoe crab, 91
hot springs, 5, 11, 12, 15, 20, 21. *See also* ocean vents
Huber, Claudia, 23
Hutchins, David, 132
hydrogen sulfide, 13, 15, 16, 23
hypobaric chambers, 108

inclination, 60, 62, 110, 111, 117, 118, 122
industry, 44, 152, 155, 156; and organic nitrogen production, 148
infertility, 133, 136
intercropping, 146
iron: atom, 5–9; biosynthesis of iron compounds, 126–127; deficiency, 31, 43, 44, 142, 149–158; at Earth's formation, *2,* 5–10, 21; in hemoglobin, 71–85; grabbers (siderophores), 31–37; and magnetic bacteria and magnetism, 49–70; and magnets in migratory animals, 127–128; and nitrogen fixation, 139–145; at origin of life, 2, 23–24, *26;* and oxygen, 2, 25, 27–29; in plants, 131, 139–145; regulation, 43–46; at sea vents, 16–18; seeding oceans with, 131–139, *136;* storage, and ferritin, 37–41; transport, via transferrin and hemoglobin, 41–43, 73–74; toxic overload of, 36, 99–101. *See also* iron atom; foods; nutrition
iron-absorption: enhancers, 153; in humans, 40–41, 73; inhibiters, 153
iron atom, *8;* in bacteria, 56–57; in cytochrome oxidase, 28; and ferritin, 37–38, 40; in hemerythrin, 91; in hemoglobin, 71–80, 87; and magnetism, 66–70; and nitrogen fixation, 140, 148; and porphyrin, 94; in seawater, 2, 27; and siderophores, 31–33; structure and function, 5–9, *8;* and thalassemia, 100; and transferrin, 41–44
iron-deficiency anemia. *See* anemia
iron oxide, 37, 56, 69, 71, 79
iron-regulatory protein, 42
iron-sulfur magnets, 63, 65
iron-sulfur protein, 23, 25, 42, 140, 159

Jacobs, Israel S., 38
Jimenez, E., 158
Johnson, Kenneth, 137
Juan de Fuca Ridge, 16, 18
Jupiter, 21

Kalmijn, Adrianus, 60
Keeton, William, 114, 115
Kemp's ridley turtle, 121–122
Kendrew, John C., 74
Klebba, Philip, 35

lactoferrin, 44, 151
lampshell worm, 91
lead, 149, 156, 158
leghemoglobin, 140, 143, 144, 145
legumes, 140, *141,* 142, *143,* 144,
    145, 146, 148, 153
Lepidochalys kempi. See Kemp's
    ridley turtle
Lichenostomus chrysops. See yellow-
    faced honeyeater
life: evolution of, *2,* 25–29, *26;* and
    iron, 7, 9, 10, 31, 38, 42, 45; at
    ocean vents, 11–17; origin of, 5,
    21, 23, 24, 49, 74; tree of, 20, *21*
light, colored, and bird orientation,
    120
lightning, 6, 10, 148
limpet, 13, 126, 127
lines of force. *See* magnetic field lines
loggerhead sea turtle, 121, 122
Lohmann, Catherine M. F., 122
Lohmann, Kenneth J., 122
Lovley, Derek R., 23, 24, 65
Lowenstam, Heinz, 126
Lozoff, Betsy, 158

magnesium, 77, 153
magnetic bacteria, 51–63, *54, 62,* 65,
    66, 106, 110, 127
magnetic domains, 38, 57–59, *58,* 67,
    68, 69, 127
magnetic exchange interaction, 69, *69*

magnetic field lines, 59–63, *61, 66,*
    106–120, *112,* 122, 124–125, 128
magnetic inclination, 122
magnetic orientation and migration,
    106–126
magnetic poles, 66; of Earth, *61, 112*
magnetism, 38, 66–69, *67, 68, 69;*
    detection mechanism, 126–128.
    See also magnetic bacteria; mag-
    netic domain; magnetic field lines;
    magnetic orientation and migra-
    tion; magnetic poles; magnetite;
    magnets
magnetite, 51, 56, 57, 59, 60, 63–66,
    64, 109, 126, 127, 128
Magnetospirillum magnetotacticum,
    53, *54,* 65
magnets, in bacteria, *54,* 55–66
malaria, 95–97, 154, 155
mammals, 23, *26,* 39, 92, 126, 127
marine animals, 92, 121–125
marine worms, 91
Martin, John, 133, 134, 135, 137, 138
Mediterranean anemia. *See* Cooley's
    anemia
Merkel, Friedrich, 107, 108
methane-making microorganisms, 19,
    20
Methanococcus jannaschii, 20
methemoglobin, 73
microorganisms: and disease, 44, 45;
    DNA of, 20; at hot springs, 11, 15,
    16; and iron, 23, 31, 32, 65, 66,
    157; and nitrogen fixation, 139,
    146; and oxygen, 25; RNA of, 19;
    in rock, 21; siderophores in, 33
midocean ridge, 11, 20
migration, 105–111, 113, 114,
    117–121, 123, 125, 129, 159
migratory unrest, 107
Miller, Stanley L., 10, 11, 20, 21, 23
mineral-eating microbe, 14
mitochondria, *26,* 72, 83, 87
molecular bonds, 7

molecules: iron, 6, 7; at origin of life, 10, 15, 20, 23, 24; oxygen, 9, 25, 27–29, 49. *See also specific elements*

Mössbauer spectroscopy, 55

Murray, John M., 154–155

muscle spindles, 128

mussel, 12, 13

myoglobin, 71, 72, 73, 74, *75,* 79, 81, 92, 145; oxygen affinity of, *83;* and oxygen delivery, *84,* 84–87

National Academy of Sciences, "Prevention of Iron Deficiency" (1998 report), 150–153

National Science Foundation, 11, 137

Neilands, J. B., 32, 33, 100

New Mexico, 21

nitrogen, 9, 10; fixation, 129, 131, 139–148, *141, 143, 144,* 159; and hemoglobin, 75–77, 81, 129; organic, 139–148

nitrogenase, 140, 143, 144, 145

nodules, plant, 142, *143,* 145

North Atlantic gyre, 122, *123*

Northern Hemisphere, *62,* 111, 118

nuclear power, 132

nucleus: and energy, 6, 132; of iron atom, 6, 7, 8; stability of, 5, 6; and magnetism, 66, 67, *67*

nutrition, and iron, 149–153

ocean vents, 12–17; and iron, 29; microorganisms at, 20, 33, 66; and origin of life, 21, 23. *See also* hot springs

oceans, 5, 20, 63; high nutrient–low chlorophyll areas of, 131–139; iron-precipitation in, *2,* 27–29; midocean ridge, 11, 20; seeding, 131–139, *136. See also* marine animals

Oncorhynchus mykiss. See rainbow trout

Oncorhynchus nerka. See sockeye salmon

Oncorhynchus tschawytscha. See chinook salmon

organisms, evolution of, 15, 20, *22,* 25

origin of life, *2, 7,* 10, 14, 21, 23, 160

Oski, Frank A., 155–156

oxidation, 9, 17, 28, 32, 39, 73, 78, 79, 80, 94

oxygen: affinity, 13, *82, 83, 84,* 87, 91, 99, 145; avoidance, 49, 63; concentration, 78, 86; delivery, *82;* docking, 79, *80;* in Earth's atmosphere, *2;* and energy, 28, 72; grabbing, 31, 36; and greenhouse effect, 132, 133, 135; in hemoglobin, 71–103, 155; in leghemoglobin, 140, *144,* 145–146; molecule, 7, 9, 21, 25, 27, 28, 29; pressure, 81–86, *82, 83, 84,* 89, 98, 99; reversible, 80. *See also* anaerobic; respiration

Pacific dolphin, 127

Passerculus sandiwichensis. See Savannah sparrow

passerines, 114

peanut worm, 91

penguin, 93

pernicious anemia, 40, 77

Perutz, Max, 74, 76, 77

photosynthesis, 1, 9, 11, 14, 15, 19, 131, 133, 140, 146

phytates, 153

phytoplankton, 132, 133, 134, 135, 137, 138, 139

phytosiderophores, 142

pied flycatcher, 117, 118

pigeon, 105, 113–120, *116,* 127, 128; and sunspots, 106

plant hemoglobin. *See* leghemoglobin

plants, 17, 23, 31, 39–40, 77; and ocean seeding, 131, 134, 137, 138;

and nitrogen fixation, 139–147, 159

poisons, 93, 146

pole, rotational, of Earth, *112*

Pollitt, Ernesto, 156–158

Popper, Sir Karl, 24

porin, 35, *36*

porphyrin, 76, 77, *77,* 80; diseases, 94, 95

precipitation, of iron, *2,* 5, 16, 18, 27, 28, 66

primordial soup theory, 10, 11, 15, 20, 21

procaryote, 17, 49, 142

protein, 35, 36, 42, 44, 45, 74; and ferritin, 37–40, *41,* 126; and hemoglobin, 73, *76, 77,* 78; myoglobin, *75,* 81, 85; and oxygen, 29, 32, 91, 92. *See also* iron-sulfur protein; iron-regulatory protein

pyrite, 16, 18, 23, 65

quantum mechanics, 7, 79

radiation, 6, 132, 148

rainbow trout, 128

rat, 126

receptors, 35; magnetic, 127, 128; transferrin, 42, *43*

red blood cells, 31, 38, 42, 87–90, 92, 94; and hemoglobin, 71–74, 78, 81–82; and iron deficiency, 151, 153; in sickle cell disease, 95–99, *98;* in thalassemia, 99–102

redox reaction, 9

reduction reaction, 9

Regulus regulus. See goldcrest

respiration, 28, 29, 86, 145

Rhizobium bacteria, 140, 141, 142–146, *143*

ribonucleic acid. *See* RNA

rice, 140, 142, 146, 148, 153

RNA, 19, 29

rock, 11, 12, 20, 27, 66, 113

rock dove, 113

Roncagliolo, Manuel, 158

rotational pole, of Earth, *112*

rust, 45; and ferritin, 38–39; and hemoglobin, 71, 73, 78, 79; ocean precipitation of, *2,* 27–29; and siderophores, 33–34

rusty liver disease, 31, 37

salmon, 105, 123, 124, 125, 127, 129

Savannah sparrow, 120

seal, 92, 142

seawater: and ocean seeding, 133, 134, 138; and origin of life, 11, 15, 16, 20, 27

seeding. *See* iron: seeding oceans with

shrimp, 13, 16

sickle cell disease, 90, 91, 95–98, *98,* 101

siderophores, 31–36, 39, 45, 47, 100. *See also* phytosiderophores

signaling, in plants, 142

silvereye, 119

sockeye salmon, 124–125

solar energy, 132

solar system, 21, 26, 27

Somali nomads, 154, 155

Southern Hemisphere, 59, 60, *61, 62,* 105, 110, 111, *112,* 119

spinach, 153

stromatolite, 21

submersible. *See Alvin*

sulfur-loving microorganisms, 16

sunlight, 9, 11, 133, 146, 148

sunspots, 106, 124

superparamagnetism, 38

Sylvia atricapilla. See European blackcap

Sylvia borin. See European garden warbler

symbiosis, of legumes and bacteria, 17, 131, 139, 140, 141, 145, 146

synthesis, of iron compounds. *See* biomineralization

teeth, 49, 51, 91, 126
thalassemia, 90–91, 95–96, 99–101, 157, 159. *See also* Cooley's anemia
thermodynamics, 7
ton B, 35, *36,* 138
transferrin, 31, 41–43, *42,* 44–46, 151, 157
tree of life, 19, 20, *22*
tube worm, *14,* 13–17
tumor, 44, 95
turtle, 103, 105, 121–124, *124,* 126, 129, 131

universe, formation of, 5, 6, 25
Urey, Harold, 10
Ustilago sphaerogena, 32

Van Andel, T. H., 12
vents, *See* ocean vents
Virginia, 20, 66
vitamins, 47, 151, 153, 154; B$_{12}$, 40, 77
Viteri, Fernando E., 150
volcanoes, 6, 11, 16, 25

Wächtershäuser, Günther, 21, 23
Walcott, Charles, 115
Walker, Michael, 128
Washington, State of 16, 18, 21, 66
whale, 105, 123, 124, 129
white-crowned sparrow, 127
WHO, 149, 154
Wiltschko, Roswitha Brill, 110–117
Wiltschko, Wolfgang, 108–117, 118, 119, 128
Woese, Carl, 17–22, 54
Wolf, A. W., 158
Wolfe, Ralph, 19, 53, 54
World Health Organization. *See* WHO

Yellowstone National Park, 11
yellow-faced honeyeater, 119
Yorke, Ellen, 128

Zonotrichia leucophrys, 127
zooplankton, 134, 137
Zosterops lateralis. See silvereye
Zugunruhe, 107

ABOUT THE AUTHORS

EUGENIE VORBURGER MIELCZAREK is Professor of Physics at George Mason University in Fairfax, Virginia. For the past twenty years, her experimental research has focused on iron in biological systems. She has been a visiting scientist at the National Institutes of Health and the Hebrew University of Jerusalem and chair of the American Institute of Physics's book publication committee. Her research has been funded by the Office of Naval Research, the National Institutes of Health, the Atomic Energy Commission, and the National Science Foundation. Mielczarek belongs to the East Coast Iron Club, a group of about forty scientists involved in studying iron in biological systems, and she has served as an adviser to National Public Radio. She is the primary editor of *Biological Physics*: Key Papers in Physics series.

SHARON BERTSCH MCGRAYNE is the author of *Nobel Prize Women in Science: Their Lives, Struggles, and Momentous Discoveries*, *365 Surprising Scientific Facts, Breakthroughs, and Discoveries*, and *Blue Genes and Polyester Plants: 365 More Surprising Scientific Facts, Breakthroughs, and Discoveries*. A former newspaper reporter and science writer and an editor for the *Encyclopaedia Britannica*, she is co-author of lengthy articles entitled "Electricity and Magnetism" and "The Atom" in the encyclopedia. She has lectured at numerous universities and scientific organizations about women in science. McGrayne lives in Seattle, Washington.